ON ANGULAR MOMENTUM

ON ANGULAR MOMENTUM

Julian Schwinger

DOVER PUBLICATIONS, INC.
Mineola, New York

Bibliographical Note

This Dover edition, first published by Dover Publications, Inc., in 2015, is an unabridged republication of the work first published for the United States Atomic Energy Commission by Technical Information Service, Oak Ridge, Tennessee, in 1952.

International Standard Book Number
ISBN-13: 978-0-486-78810-4
ISBN-10: 0-486-78810-5

Manufactured in the United States by Courier Corporation
78810501   2015
www.doverpublications.com

ON ANGULAR MOMENTUM

ON ANGULAR MOMENTUM

Julian Schwinger

The commutation relations of an arbitrary angular momentum vector can be reduced to those of the harmonic oscillator. This provides a powerful method for constructing and developing the properties of angular momentum eigenvectors. In this paper many known theorems are derived in this way, and some new results obtained. Among the topics treated are the properties of the rotation matrices; the addition of two, three, and four angular momenta; and the theory of tensor operators.

1. Introduction

One of the methods of treating a general angular momentum in quantum mechanics is to regard it as the superposition of a number of elementary "spins," or angular momenta with $j = \frac{1}{2}$. Such a spin assembly, considered as a Bose-Einstein system, can be usefully discussed by the method of second quantization. We shall see that this procedure unites the compact symbolism of the group theoretical approach with the explicit operator techniques of quantum mechanics.

We introduce spin creation and annihilation operators associated with a given spatial reference system, $a_\zeta^+ = (a_+^+, a_-^+)$ and $a_\zeta = (a_+, a_-)$, which satisfy

$$[a_\zeta, a_{\zeta'}] = 0, \quad [a_\zeta^+, a_{\zeta'}^+] = 0,$$
$$[a_\zeta, a_{\zeta'}^+] = \delta_{\zeta\zeta'}. \tag{1.1}$$

The number of spins and the resultant angular momentum are then given by

$$n = \sum_\zeta a_\zeta^+ a_\zeta = n_+ + n_-,$$
$$\mathbf{J} = \sum_{\zeta,\zeta'} a_\zeta^+ (\zeta|\tfrac{1}{2}\boldsymbol{\sigma}|\zeta') a_{\zeta'}. \tag{1.2}$$

With the conventional matrix representation for $\boldsymbol{\sigma}$, the components of $\mathbf{J}$ appear as

$$J_+ = J_1 + iJ_2 = a_+^+ a_-, \quad J_- = J_1 - iJ_2 = a_-^+ a_+,$$
$$J_3 = \tfrac{1}{2}(a_+^+ a_+ - a_-^+ a_-) = \tfrac{1}{2}(n_+ - n_-). \tag{1.3}$$

Of course, this realization of the angular momentum commutation properties in terms of those of harmonic oscillators can be introduced without explicit reference to the composition of spins.

To evaluate the square of the total angular momentum

$$\mathbf{J}^2 = \sum_{\zeta\zeta'\zeta''\zeta'''} a_\zeta^+ a_{\zeta'} a_{\zeta''}^+ a_{\zeta'''} (\zeta|\tfrac{1}{2}\boldsymbol{\sigma}|\zeta') \cdot (\zeta''|\tfrac{1}{2}\boldsymbol{\sigma}|\zeta'''), \tag{1.4}$$

we employ the matrix elements of the spin permutation operator

$$P^{(12)} = \tfrac{1}{2}(1 + \boldsymbol{\sigma}^{(1)} \cdot \boldsymbol{\sigma}^{(2)}). \tag{1.5}$$

Thus

$$(\zeta|\boldsymbol{\sigma}|\zeta') \cdot (\zeta''|\boldsymbol{\sigma}|\zeta''') = 2\delta_{\zeta\zeta'''}\delta_{\zeta''\zeta'} - \delta_{\zeta\zeta'}\delta_{\zeta''\zeta'''}, \tag{1.6}$$

and

$$\mathbf{J}^2 = \tfrac{1}{2}\sum_{\zeta\zeta''} a_\zeta^+ a_{\zeta''} a_{\zeta''}^+ a_\zeta - \tfrac{1}{4}n^2. \tag{1.7}$$

According to the commutation relations (1.1),

$$\sum_{\zeta,\zeta'} a_\zeta^+ a_{\zeta'} a_{\zeta'}^+ a_\zeta = \sum_\zeta a_\zeta^+ (n+2) a_\zeta = n(n+1), \tag{1.8}$$

whence

$$\mathbf{J}^2 = \tfrac{1}{2}n(\tfrac{1}{2}n + 1); \tag{1.9}$$

a given number of spins, $n = 0, 1, 2, \ldots$, possesses a definite angular momentum quantum number,

$$j = \tfrac{1}{2}n = 0, \tfrac{1}{2}, 1, \ldots. \tag{1.10}$$

We further note that, according to (1.3), a state with a fixed number of positive and negative spins also has a definite magnetic quantum number,

$$m = \tfrac{1}{2}(n_+ - n_-), \qquad j = \tfrac{1}{2}(n_+ + n_-). \tag{1.11}$$

Therefore, from the eigenvector of a state with prescribed occupation numbers,

$$\Psi(n_+ n_-) = \frac{(a_+^+)^{n_+}}{(n_+!)^{1/2}} \frac{(a_-^+)^{n_-}}{(n_-!)^{1/2}} \Psi_0, \tag{1.12}$$

$$a_\pm \Psi_0 = 0,$$

we obtain the angular momentum eigenvector[1]

$$\Psi(jm) = \frac{(a_+^+)^{j+m}(a_-^+)^{j-m}}{[(j+m)!(j-m)!]^{1/2}} \Psi_0. \tag{1.13}$$

Familiar as a symbolic expression of the transformation properties of angular momentum eigenvectors[2], this form is here a precise operator construction of the eigenvector.

---

[1] A direct proof is given in Appendix A.

[2] See, for example, H. Weyl, "The Theory of Groups and Quantum Mechanics" (E. P. Dutton and Company, Inc., New York, 1931), p. 189.

On multiplying (1.13) with an analogous monomial constructed from the components of the arbitrary spinor $x_\zeta = (x_+, x_-)$

$$\phi_{jm}(x) = \frac{x_+^{j+m} x_-^{j-m}}{[(j+m)!(j-m)!]^{1/2}} \tag{1.14}$$

we obtain, after summation with respect to m, and then with respect to j,

$$\sum_{m=-j}^{j} \phi_{jm}(x)\Psi(jm) = \frac{(xa^+)^{2j}}{(2j)!} \Psi_0, \tag{1.15}$$

and

$$\sum_{jm} \phi_{jm}(x)\Psi(jm) = e^{(xa^+)}\Psi_0, \tag{1.16}$$

in which we have written

$$(xa^+) = \sum_\zeta x_\zeta a_\zeta^+. \tag{1.17}$$

To illustrate the utility of (1.16), conceived of as an eigenvector generating function, we shall verify the orthogonality and normalization of the eigenvectors (1.13). Consider, then,

$$(e^{(xa^+)}\Psi_0, e^{(ya^+)}\Psi_0) = \sum \phi_{jm}(x^*)(\Psi(jm), \Psi(j'm'))\phi_{j'm'}(y)$$
$$= (\Psi_0, e^{(x^*a)}e^{(ya^+)}\Psi_0). \tag{1.18}$$

According to the commutation relations (1.1), and $a_\zeta \Psi_0 = 0$, we have

$$a_\zeta f(a^+)\Psi_0 = \left(\frac{\partial f(a^+)}{\partial a_\zeta^+}\right)\Psi_0, \tag{1.19}$$

whence

$$(\Psi_0, e^{(x^*a)} e^{(ya^+)}\Psi_0) = e^{(x^*y)}(e^{(y^*a)}\Psi_0, \Psi_0) = e^{(x^*y)}$$
$$= \sum_{jm} \phi_{jm}(x^*)\phi_{jm}(y). \tag{1.20}$$

We have thus proved that

$$(\Psi(jm), \Psi(j'm')) = \delta_{jj'} \delta_{mm'}. \tag{1.21}$$

As a second elementary example, we shall obtain the matrix elements of powers of $J_\pm$ by considering the effect of the operators $e^{\lambda J_\pm}$ on (1.16). We have

$$\sum_{jm} \phi_{jm}(x)e^{\lambda J_+}\Psi(jm) = e^{\lambda a_+^+ a_-} e^{(xa^+)}\Psi_0 = e^{\lambda x_- a_+^+} e^{(xa^+)}\Psi_0$$
$$= e^{(x_+ + \lambda x_-)a_+^+ + x_- a_-^+}\Psi_0 \tag{1.22}$$
$$= \sum_{jm} \phi_{jm}(x_+ + \lambda x_-, x_-)\Psi(jm),$$

and therefore
$$\sum_{j'm'} (jm|e^{\lambda J_+}|j'm')\phi_{j'm'}(x) = \phi_{jm}(x_+ + \lambda x_-, x_-), \quad (1.23)$$

which, on expansion, yields the nonvanishing matrix element
$$(jm|J_+^{m-m'}|jm') = \left[\frac{(j+m)!}{(j+m')!}\frac{(j-m')!}{(j-m)!}\right]^{1/2}, \quad m-m' > 0. \quad (1.24)$$

Similarly
$$\sum_{j'm'} (jm|e^{\lambda J_-}|j'm')\phi_{j'm'}(x) = \phi_{jm}(x_+, x_- + \lambda x_+), \quad (1.25)$$

and
$$(jm|J_-^{m'-m}|jm') = \left[\frac{(j+m')!}{(j+m)!}\frac{(j-m)!}{(j-m')!}\right]^{1/2}, \quad m'-m > 0. \quad (1.26)$$

A particular consequence of (1.24) and (1.26) is
$$\Psi(jm) = \left[\frac{1}{(2j)!}\frac{(j-m)!}{(j+m)!}\right]^{1/2} \cdot J_+^{j+m}\Psi(j,-j)$$
$$= \left[\frac{1}{(2j)!}\frac{(j+m)!}{(j-m)!}\right]^{1/2} \cdot J_-^{j-m}\Psi(jj), \quad (1.27)$$

which details the construction of an arbitrary eigenvector from those possessing the maximum values of $|m|$ compatible with a given j.

It is also possible to exhibit an operator which permits the construction of an arbitrary eigenvector from that possessing the minimum value of j compatible with a given m. Indeed, (1.13), written in the form
$$\Psi(jm) = \frac{(a_+^+ a_-^+)^{j-|m|}}{[(j+|m|)!(j-|m|)!]^{1/2}} (a_+^+)^{|m|+m}(a_-^+)^{|m|-m}\Psi_0, \quad (1.28)$$

states that
$$\Psi(jm) = \left[\frac{(2|m|)!}{(j+|m|)!(j-|m|)!}\right]^{1/2} K_+^{j-|m|} \Psi(|m|,m), \quad (1.29)$$

where K_+ and two associated operators are defined by
$$K_+ = a_+^+ a_-^+, \quad K_- = a_+ a_-, \quad (1.30)$$
$$K_3 = \tfrac{1}{2}(n_+ + n_- + 1).$$

It is easily seen that
$$[J_3, K_\pm] = [J_3, K_3] = 0, \quad (1.31)$$
and that
$$[K_3, K_+] = K_+, \quad [K_3, K_-] = -K_-,$$
$$[K_+, K_-] = -2K_3. \quad (1.32)$$

The latter are analogous to the commutation properties of J, save for the algebraic sign of the commutator $[K_+, K_-]$. In keeping with this qualified analogy we also have

$$J_3^2 - \tfrac{1}{4} = K_3(K_3 - 1) - K_+ K_- = K_3(K_3 + 1) - K_- K_+ \qquad (1.33)$$

as compared with

$$J^2 = J_3(J_3 - 1) + J_+ J_- = J_3(J_3 + 1) + J_- J_+. \qquad (1.34)$$

Noting that the eigenvalue of K_3 is $j + \tfrac{1}{2}$, we see that the roles of j and m are essentially interchanged in K. The hyperbolic nature of the space in which the latter operates is thus related to the restriction $|m| \le j$.

If (1.29) is multiplied by a similar numerical quantity, and then summed with respect to j, one obtains

$$\sum_{j=|m|}^{\infty} \left[\frac{(2|m|)!}{(j+|m|)!(j-|m|)!} \right]^{1/2} \lambda^{j-|m|} \Psi^\circ(jm) = F_{2|m|}(\lambda K_+) \Psi^\circ(|m|, m), \qquad (1.35)$$

where

$$F_r(z) = r! z^{-r/2} I_r(2z^{1/2}) = \sum_{n=0}^{\infty} \frac{r!}{n!(r+n)!} z^n, \qquad (1.36)$$

and I_r is the cylinder function of imaginary argument. A simpler generating function is given by

$$\sum_j \left[\frac{1}{(2|m|)!} \frac{(j+|m|)!}{(j-|m|)!} \right]^{1/2} \lambda^{j-|m|} \Psi^\circ(jm) = e^{\lambda K_+} \Psi^\circ(|m|, m). \qquad (1.37)$$

2. Rotations

A significant interpretation is obtained for (1.15) by introducing the operators

$$a'^+_+ = (xa^+), \qquad a'_+ = (x^*a), \qquad (2.1)$$

$$a'^+_- = [x^* a^+], \qquad a'_- = [xa],$$

where

$$[xy] = x_+ y_- - x_- y_+. \qquad (2.2)$$

With the restriction

$$(x^*x) = 1, \qquad (2.3)$$

these operators also obey the commutation relations (1.1), and must therefore constitute spin creation and annihilation operators associated with an altered spatial reference system. Accordingly, (1.15) can be viewed as the expression of the state $m = j$, in a rotated coordinate system, as a linear combination of the eigenvectors in a fixed coordinate system,

$$\Psi''(jj) = \frac{(a'^+_+)^{2j}}{((2j)!)^{1/2}} \Psi_0 = ((2j)!)^{1/2} \sum_{m=-j}^{j} \phi_{jm}(x)\Psi(jm). \qquad (2.4)$$

The unitary nature of this transformation is here easily verified,

$$(2j)! \sum_m \phi_{jm}(x^*)\phi_{jm}(x) = (x^*x)^{2j} = 1. \qquad (2.5)$$

In general

$$\Psi''(jm') = \phi_{jm'}(a'^+)\Psi_0 = \sum_m \Psi(jm) U^{(j)}_{mm'}, \qquad (2.6)$$

where the coefficients are to be inferred from

$$\sum_m \phi_{jm}(a^+) U^{(j)}_{mm'} = \phi_{jm'}(x_+a^+_+ + x_-a^+_-, -x^*_-a^+_+ + x^*_+a^+_-). \qquad (2.7)$$

It is useful to introduce the unitary operator that generates $\Psi''(jm')$ from $\Psi(jm)$,

$$\Psi''(jm') = U\Psi(jm), \qquad (2.8)$$

which permits an alternative construction of the coefficients in (2.6),

$$U^{(j)}_{mm'} = (jm|U|jm'). \qquad (2.9)$$

In terms of the successive rotations characterized by Eulerian angles, ϕ, θ, ψ, U is given explicitly by

$$U = e^{-i\psi J_3''} e^{-i\theta J_2'} e^{-i\phi J_3}, \qquad (2.10)$$

where

$$J_2' = e^{-i\phi J_3} J_2 e^{i\phi J_3},$$

$$J_3'' = e^{-i\theta J_2'} J_3' e^{i\theta J_2'} \qquad (2.11)$$

are the operators appropriate to the coordinate systems produced by the previous rotations. The resulting expression for $U(\phi\theta\psi)$ is

$$U = e^{-i\phi J_3} e^{-i\theta J_2} e^{-i\psi J_3}, \qquad U^{-1} = e^{i\psi J_3} e^{i\theta J_2} e^{i\phi J_3}. \qquad (2.12)$$

The angular momentum operators associated with the new coordinate system,

$$J' = UJU^{-1}, \qquad (2.13)$$

can be constructed from the transformed creation and annihilation operators,

$$a'^+_+ = Ua^+_+U^{-1} = e^{-(i/2)(\psi+\phi)} \cos \tfrac{1}{2}\theta \, a^+_+ + e^{-(i/2)(\psi-\phi)} \sin \tfrac{1}{2}\theta \, a^+_- \qquad (2.14)$$

$$a'^+_- = Ua^+_-U^{-1} = -e^{-(i/2)(\psi-\phi)} \sin \tfrac{1}{2}\theta \, a^+_+ + e^{(i/2)(\psi+\phi)} \cos \tfrac{1}{2}\theta \, a^+_-.$$

In evaluating (2.14), we have made use of the relations

$$e^{-i\psi J_3} a^+_\pm e^{i\psi J_3} = e^{\mp (i/2)\psi} a^+_\pm,$$

$$e^{-i\theta J_2} a^+_\pm e^{i\theta J_2} = \cos \tfrac{1}{2}\theta\, a^+_\pm \pm \sin \tfrac{1}{2}\theta\, a^+_\mp, \qquad (2.15)$$

of which the former follows immediately from the significance of $a^+_\pm$ as a positive (negative) spin creation operator, while the latter may be verified by differentiation with respect to θ, in conjunction with the commutation relations

$$[a^+_\pm, J_2] = \mp\,(i/2)\,a^+_\mp. \qquad (2.16)$$

The form of (2.14) is in agreement with (2.1) and (2.3), where

$$x_+ = e^{-(i/2)(\psi+\phi)} \cos \tfrac{1}{2}\theta, \quad x_- = e^{-(i/2)(\psi-\phi)} \sin \tfrac{1}{2}\theta. \qquad (2.17)$$

To construct the matrix of U, we consider

$$(e^{(xa^+)}\Psi_0,\, Ue^{(ya^+)}\Psi_0) = \sum_{jm} \phi_{jm}(x^*) U^{(j)}_{mm'} \phi_{jm'}(y)$$

$$= (\Psi_0,\, e^{(x^*a)} e^{(ya'^+)}\Psi_0), \qquad (2.18)$$

in which the a'^+ are the operators (2.14). On writing

$$(ya'^+) = (a^+ uy) \qquad (2.19)$$

where u is the matrix

$$u = \begin{pmatrix} e^{-(i/2)(\phi+\psi)} \cos \tfrac{1}{2}\theta, & -e^{-(i/2)(\phi-\psi)} \sin \tfrac{1}{2}\theta \\ e^{-(i/2)(\phi-\psi)} \sin \tfrac{1}{2}\theta, & e^{(i/2)(\phi+\psi)} \cos \tfrac{1}{2}\theta \end{pmatrix} \qquad (2.20)$$

we immediately obtain

$$\sum_{jm} \phi_{jm}(x^*) U^{(j)}_{mm'} \phi_{jm'}(y) = e^{(x^* uy)}. \qquad (2.21)$$

Since (2.12) implies that

$$U^{(j)}_{mm'}(\phi\theta\psi) = e^{-im\phi}\, U^{(j)}_{mm'}(\theta) e^{-im'\psi}, \qquad (2.22)$$

where

$$U^{(j)}_{mm'}(\theta) = (jm|e^{-i\theta J_2}|jm'), \qquad (2.23)$$

we may simplify (2.21) by placing $\phi = \psi = 0$, thereby obtaining

$$\sum_{jm} \phi_{jm}(x^*) U^{(j)}_{mm'}(\theta) \phi_{jm'}(y) = \exp\{\cos \tfrac{1}{2}\theta (x^*y) - \sin \tfrac{1}{2}\theta [x^*y]\}. \qquad (2.24)$$

The matrix u is unitary and unimodular, that is, possesses a unit determinant. Its representation in terms of spin matrices has, as it must, the form of (2.12),

$$u = e^{-(i/2)\phi\sigma_3}\, e^{-(i/2)\theta\sigma_2}\, e^{-(i/2)\psi\sigma_3}. \qquad (2.25)$$

Any such unitary matrix can be presented as
$$u = e^{-i\mathcal{H}} \tag{2.26}$$
where $\mathcal{H}$ is a Hermitian matrix. Since
$$\det u = e^{-i\,\mathrm{tr}\,\mathcal{H}}, \tag{2.27}$$
$\mathcal{H}$ must be a traceless Hermitian matrix and, accordingly, is a linear combination of the spin matrices, with real coefficients. Hence u can be written as
$$u = e^{-(i/2)\gamma \mathbf{n}\cdot\boldsymbol{\sigma}} \tag{2.28}$$
where $\mathbf{n}$ is a unit vector, specified by two angles, α and β. The fact that (2.28) is the matrix describing a rotation through the angle γ about the axis $\mathbf{n}$ affirms the well-known equivalence between an arbitrary rotation and a simple rotation about a suitably chosen axis. The rotation angle γ is easily obtained by comparing the trace of u, in its two versions,
$$\tfrac{1}{2}\mathrm{tr}\, u = \cos \tfrac{1}{2}\gamma = \cos \tfrac{1}{2}\theta \cos \tfrac{1}{2}(\phi+\psi). \tag{2.29}$$

More generally, the trace of U for a given j depends only upon the rotation angle γ. We define[3]
$$\chi^{(j)} = \sum_{m=-j}^{j} U_{mm}^{(j)} = \mathrm{tr}\, P_j U, \tag{2.30}$$
in which P_j is the projection operator for the states with quantum number j. If we remark that U must also have the form of (2.28),
$$U = e^{-i\gamma \mathbf{n}\cdot\mathbf{J}} \tag{2.31}$$
we immediately obtain
$$\chi^{(j)} = \sum_{m=-j}^{j} e^{-im\gamma} = \frac{\sin(j+\tfrac{1}{2})\gamma}{\sin \tfrac{1}{2}\gamma}. \tag{2.32}$$

However, we can also derive this directly from the generating function (2.21).

For simplicity we shall assume the reference system to be so chosen that u is a diagonal matrix, with eigenvalues $e^{\pm(i/2)\gamma}$. We replace x_ζ^* with $t(\partial/\partial y_\zeta)$ and evaluate the derivatives at $y_\zeta = 0$. According to
$$\phi_{jm}(\partial/\partial y)\phi_{jm'}(y)]_{y_\zeta=0} = \delta_{m,m'}, \tag{2.33}$$
we then have
$$\sum_j t^{2j}\chi^{(j)} = \exp\left(te^{-(i/2)\gamma}\frac{\partial}{\partial y_+}; y_+\right)\cdot\exp\left(te^{(i/2)\gamma}\frac{\partial}{\partial y_-}; y_-\right)\bigg]_{y_\zeta=0} \tag{2.34}$$
in which the notation reflects the necessity of placing the derivatives to the left of the powers of y_ζ. Now

---
[3] This trace is the character of group theory.

$$\exp\left(\lambda\frac{\partial}{\partial y};y\right)=\sum_{n=0}^{\infty}\frac{\lambda^n}{n!}\left(\frac{\partial}{\partial y}\right)^n y^n=\sum_{n=0}^{\infty}\lambda^n=\frac{1}{1-\lambda}, \tag{2.35}$$

and therefore

$$\sum_j t^{2j}\chi^{(j)}=\frac{1}{1-t\cdot\exp\left(-\frac{i}{2}\gamma\right)}\frac{1}{1-t\cdot\exp\left(\frac{i}{2}\gamma\right)}$$

$$=\frac{1}{1-2t\cdot\cos\tfrac{1}{2}\gamma+t^2}, \tag{2.36}$$

which is a generating function for the $\chi^{(j)}$. On writing

$$\frac{1}{1-t\cdot\exp\left(-\frac{i}{2}\gamma\right)}\frac{1}{1-t\cdot\exp\left(\frac{i}{2}\gamma\right)}$$

$$=\frac{1}{2it\cdot\sin\tfrac{1}{2}\gamma}\left[\frac{1}{1-t\cdot\exp\left(\frac{i}{2}\gamma\right)}-\frac{1}{1-t\cdot\exp\left(-\frac{i}{2}\gamma\right)}\right], \tag{2.37}$$

and expanding in powers of t, one obtains

$$\chi^{(j)}(\gamma)=\frac{\sin(j+\tfrac{1}{2})\gamma}{\sin\tfrac{1}{2}\gamma}. \tag{2.38}$$

Symmetry properties of $U^{(j)}_{mm'}(\phi\theta\psi)$ are easily inferred from (2.21). According to the invariance of (x^*uy) under the substitutions $\phi\leftrightarrow\psi+\pi$, $x^*\leftrightarrow y$, and $\phi\to\phi-\pi$, $\theta\to\pi-\theta$, $\psi\to-\psi$, $y_\pm\to iy_\mp$, we have

$$U^{(j)}_{mm'}(\phi\theta\psi)=U^{(j)}_{m'm}(\psi+\pi,\theta,\phi-\pi)=i^{2j}U^{(j)}_{m,-m'}(\phi-\pi,\pi-\theta,-\psi). \tag{2.39}$$

Among the additional equivalent forms produced by successive application of these transformations are

$$i^{2j}U^{(j)}_{-mm'}(-\phi,\pi-\theta,\psi+\pi)=U^{(j)}_{-m-m'}(\pi-\phi,\theta,-\pi-\psi)$$

$$=U^{(j)}_{-m'-m}(-\psi,\theta,-\phi). \tag{2.40}$$

We also note that

$$U^{(j)*}_{mm'}(\phi\theta\psi)=U^{(j)}_{mm'}(-\phi,\theta,-\psi)=U^{(j)}_{-m-m'}(\phi+\pi,\theta,\psi-\pi). \tag{2.41}$$

On removing the angles ϕ and ψ with the aid of (2.22), we find that the content of (2.39) and (2.40) is

$$U^{(j)}_{mm'}(\theta)=(-1)^{j-m}U^{(j)}_{m-m'}(\pi-\theta)=(-1)^{j-m'}U^{(j)}_{-mm'}(\pi-\theta)$$

$$=(-1)^{m-m'}U^{(j)}_{-m-m'}(\theta)=(-1)^{m-m'}U^{(j)}_{m'm}(\theta)=U^{(j)}_{-m'-m}(\theta). \tag{2.42}$$

In view of these relations, it is sufficient to exhibit $U_{mm'}^{(j)}(\theta)$ for non-negative values of m and m'.

On expanding the generating function (2.24) in terms of $\phi_{jm}(x^*)$, or of $\phi_{jm'}(y)$, we obtain the equivalent expressions

$$\sum_{m'} U_{mm'}^{(j)}(\theta) \phi_{jm'}(y) = \phi_{jm}(\cos \tfrac{1}{2}\theta\, y_+ - \sin \tfrac{1}{2}\theta\, y_-, \sin \tfrac{1}{2}\theta\, y_+ + \cos \tfrac{1}{2}\theta\, y_-), \quad (2.43a)$$

$$\sum_{m} \phi_{jm}(x^*) U_{mm'}^{(j)}(\theta) = \phi_{jm'}(\cos \tfrac{1}{2}\theta\, x_+^* + \sin \tfrac{1}{2}\theta\, x_-^*, -\sin \tfrac{1}{2}\theta\, x_+^* + \cos \tfrac{1}{2}\theta x_-^*), \quad (2.43b)$$

of which the latter is the counterpart of (2.7). As a convenient means of constructing $U_{mm'}^{(j)}(\theta)$, we place

$$x_+^* = \sin \tfrac{1}{2}\theta \cos \tfrac{1}{2}\theta, \qquad x_-^* = t - \cos^2 \tfrac{1}{2}\theta,$$

so that (2.43b) reads

$$\sum_m \frac{(\sin \tfrac{1}{2}\theta \cos \tfrac{1}{2}\theta)^{j+m}}{[(j+m)!(j-m)!]^{1/2}} (t - \cos^2 \tfrac{1}{2}\theta)^{j-m} U_{mm'}^{(j)}(\theta)$$

$$= (-1)^{j-m'} \left[\frac{(\sin \tfrac{1}{2}\theta)^{j+m'}(\cos \tfrac{1}{2}\theta)^{j-m'}}{[(j+m')!(j-m')!]^{1/2}} \right] t^{j+m'}(1-t)^{j-m'}. \quad (2.44)$$

Thus

$$U_{mm'}^{(j)}(\theta) = (-1)^{j-m'} \left[\frac{(j+m)!}{(j-m)!} \frac{1}{(j+m')!(j-m')!} \right]^{1/2}$$

$$\cdot [(\sin \tfrac{1}{2}\theta)^{-m+m'} (\cos \tfrac{1}{2}\theta)^{-m-m'}]$$

$$\cdot \left[\left(\frac{d}{dt}\right)^{j-m} t^{j+m'}(1-t)^{j-m'} \right]_{t=\cos^2 \tfrac{1}{2}\theta}. \quad (2.45)$$

The structure of the right side will be recognized as that of the Jacobi polynomial,

$$\mathscr{F}_n(a, b; t) = F(-n, a+n, b; t) = \frac{(b-1)!}{(b+n-1)!} t^{1-b}(1-t)^{b-a}$$

$$\cdot \left(\frac{d}{dt}\right)^n t^{b+n-1}(1-t)^{a-b+n}, \quad (2.46)$$

whence[4]

$$U_{mm'}^{(j)}(\theta) = \frac{(-1)^{j-m'}}{(m+m')!} \left[\frac{(j+m)!}{(j-m)!} \frac{(j+m')!}{(j-m')!} \right]^{1/2} (\sin \tfrac{1}{2}\theta)^{m-m'} (\cos \tfrac{1}{2}\theta)^{m+m'}$$

$$\cdot \mathscr{F}_{j-m}(2m+1, m+m'+1; \cos^2 \tfrac{1}{2}\theta). \quad (2.47)$$

---

[4] This is equivalent to the result obtained by P. Güttinger, Z. Phys. 73, 169 (1931).

Other forms can be obtained from (2.43), corresponding to the variety of transformations permissible to hypergeometric functions. Thus the known relation

$$F(a, b, c; x) = (1-x)^{-a} F\left(a, c-b, c; -\frac{x}{1-x}\right), \quad (2.48)$$

applied to (2.47), gives

$$U_{mm'}^{(j)}(\theta) = \frac{(-1)^{j-m'}}{(m+m')!} \left[\frac{(j+m)!\,(j+m')!}{(j-m)!\,(j-m')!}\right]^{1/2} (\sin \tfrac{1}{2}\theta)^{2j} (\cot \tfrac{1}{2}\theta)^{m+m'}$$
$$\cdot F(m-j, m'-j, m+m'+1; -\cot^2 \tfrac{1}{2}\theta). \quad (2.49)$$

Another aspect of reference system transformation is best discussed in terms of

$$U_{mm'}^{(j)*}(\phi\theta\psi) = e^{im\phi} U_{mm'}^{(j)}(\theta) e^{im'\psi} = (jm'|U^{-1}|jm). \quad (2.50)$$

This quantity is the transformation function

$$(\Psi'(jm'), \Psi(jm)) = (\omega, jm'|jm), \quad (2.51)$$

in which we have used ω to designate collectively the angles $\phi\theta\psi$, relating the new reference system to the fixed one. We shall be interested in the differential characterization of this transformation function, in its dependence upon the Eulerian angles. Now

$$\frac{1}{i}\frac{\partial}{\partial \phi} U^{-1} = U^{-1} J_3$$

$$\frac{1}{i}\frac{\partial}{\partial \psi} U^{-1} = J_3 U^{-1} = U^{-1} J_3' \quad (2.52)$$

$$\frac{1}{i}\frac{\partial}{\partial \theta} U^{-1} = U^{-1} e^{-i\phi J_3} J_2 e^{i\phi J_3} = U^{-1} J_\theta,$$

where

$$J_3' = J_3 \cos\theta + \tfrac{1}{2} \sin\theta (J_+ e^{-i\phi} + J_- e^{i\phi}),$$
$$J_\theta = \frac{1}{2i}(J_+ e^{-i\phi} - J_- e^{i\phi}), \quad (2.53)$$

and, therefore

$$\frac{1}{i}\frac{\partial}{\partial \phi}(\omega| \ ) = (\omega|J_3| \ )$$

$$e^{i\phi}\left[\frac{\partial}{\partial \theta} + \frac{1}{\sin\theta}\left(\frac{1}{i}\frac{\partial}{\partial \psi} - \cos\theta \frac{1}{i}\frac{\partial}{\partial \phi}\right)\right](\omega| \ ) = (\omega|J_+| \ ) \quad (2.54)$$

$$e^{-i\phi}\left[-\frac{\partial}{\partial \theta} + \frac{1}{\sin\theta}\left(\frac{1}{i}\frac{\partial}{\partial \psi} - \cos\theta \frac{1}{i}\frac{\partial}{\partial \phi}\right)\right](\omega| \ ) = (\omega|J_-| \ ).$$

This is a differential operator representation of an arbitrary angular momentum vector. The familiar differential operators associated with an orbital angular momentum emerge if the transformation function is independent of ψ. Since this corresponds to $m' = 0$, the quantum number j must then be an integer.[5]

The differential operators (2.54) are well-known in connection with angular momentum of a rigid body, and, accordingly, the eigenvalue equation for J^2 in this representation will be identical with the symmetrical top wave equation. To construct this equation directly, we remark that

$$J^2 = J_3^2 + (\tfrac{1}{2}J_+ e^{-i\phi} + \tfrac{1}{2}J_- e^{i\phi})^2 - (\tfrac{1}{2}J_+ e^{-i\phi} - \tfrac{1}{2}J_- e^{i\phi})^2$$

$$= J_3^2 + \left[\frac{J_3' - J_3 \cos\theta}{\sin\theta}\right]^2 + J_\theta^2 \tag{2.55}$$

$$= \frac{J_3'^2 - 2J_3'J_3\cos\theta + J_3^2}{\sin^2\theta} + J_\theta^2 + \cot\theta \frac{1}{i}J_\theta,$$

since

$$[J_3', J_3] = \sin\theta \frac{1}{i} J_\theta. \tag{2.56}$$

On referring to (2.52), we immediately obtain

$$-\left[\frac{\partial^2}{\partial\theta^2} + \cot\theta \frac{\partial}{\partial\theta} + \frac{1}{\sin^2\theta}\left(\frac{\partial^2}{\partial\psi^2} - 2\cos\theta \frac{\partial}{\partial\psi}\frac{\partial}{\partial\phi} + \frac{\partial^2}{\partial\phi^2}\right)\right] U^{-1} = U^{-1} J^2, \tag{2.57}$$

and the analogous differential equation for $(\omega|\ )$, including the eigenvalue equation

$$\left[\frac{\partial^2}{\partial\theta^2} + \cot\theta \frac{\partial}{\partial\theta} + j(j+1) - \frac{m^2 - 2mm'\cos\theta + m'^2}{\sin^2\theta}\right](\omega, jm'|jm) = 0. \tag{2.58}$$

An integral theorem concerning the angular dependence of U, or U^{-1}, is stated by

$$\int U\, d\omega = P_0, \tag{2.59}$$

where P_0 is the projection operator for the state $j = 0$, and

$$d\omega = \tfrac{1}{2}\sin\theta \cdot d\theta \cdot \frac{1}{4\pi} d\phi \cdot \frac{1}{4\pi} d\psi, \tag{2.60}$$

$$\int d\omega = 1.$$

---

[5] The fact that the general differential operators (2.54) admit half-integral values of j has been noticed by F. Bopp and R. Haag, Z. Naturforsch. **5a**, 644 (1950).

The integration domain is here understood to be

$$0 \leq \phi < 4\pi, \quad 0 \leq \psi < 4\pi, \quad 0 \leq \theta \leq \pi. \tag{2.61}$$

To prove this theorem we subject (2.57) to the angular integrations contained in $d\omega$. In virtue of the periodicity possessed by U^{-1} over 4π intervals of ϕ and ψ, we obtain

$$\int U^{-1} d\omega \, J^2 = -\tfrac{1}{2} \left[\sin\theta \, \frac{\partial}{\partial \theta} \int U^{-1} \frac{d\phi}{4\pi} \frac{d\psi}{4\pi} \right]_{\theta=0}^{\pi} = 0. \tag{2.62}$$

This result asserts the vanishing of $\int U^{-1} d\omega$, and the Hermitian conjugate $\int U \, d\omega$, except for the state with $j=0$. The fact that the rotation operator U reduces to unity for this spherically symmetrical state completes the proof of (2.59). We shall defer application of this theorem to the next section.

3. Addition of Two Angular Momenta

Two kinematically independent angular momenta, $\mathbf{J}_1$ and $\mathbf{J}_2$, can be expressed by

$$\mathbf{J}_1 = \sum_{\zeta,\zeta'} a_\zeta^+ (\zeta|\tfrac{1}{2}\boldsymbol{\sigma}|\zeta') a_{\zeta'},$$

$$\mathbf{J}_2 = \sum_{\zeta,\zeta'} b_\zeta^+ (\zeta|\tfrac{1}{2}\boldsymbol{\sigma}|\zeta') b_{\zeta'}, \tag{3.1}$$

where the a and b operators individually obey (1.1), but are mutually commutative. In studying the eigenvectors of the total angular momentum,

$$\mathbf{J} = \mathbf{J}_1 + \mathbf{J}_2, \tag{3.2}$$

the following scalar operators play an important role:

$$\mathscr{J}_+ = (a^+ b), \quad \mathscr{J}_- = (b^+ a),$$

$$\mathscr{J}_3 = \tfrac{1}{2}[(a^+ a) - (b^+ b)] = \tfrac{1}{2}(n_1 - n_2), \tag{3.3}$$

and

$$\mathscr{K}_+ = [a^+ b^+], \quad \mathscr{K}_- = [ab],$$

$$\mathscr{K}_3 = \tfrac{1}{2}[(a^+ a) + (b^+ b)] + 1 = \tfrac{1}{2}n + 1. \tag{3.4}$$

As one can easily verify by direct calculation, the operators $\mathscr{J}$ and $\mathscr{K}$ commute with each other (as well as with J), and obey

$$[\mathscr{J}_3, \mathscr{J}_\pm] = \pm \mathscr{J}_\pm, \quad [\mathscr{J}_+, \mathscr{J}_-] = 2\mathscr{J}_3,$$

$$[\mathscr{K}_3, \mathscr{K}_\pm] = \pm \mathscr{K}_\pm, \quad [\mathscr{K}_+, \mathscr{K}_-] = -2\mathscr{K}_3. \tag{3.5}$$

It will be noted that the commutation properties of the $\mathscr{J}$ operators are those of a conventional angular momentum, while the $\mathscr{K}$ operators are analogous to the hyperbolic angular momentum K, which was discussed in the first section. We shall denote the eigenvalues of $\mathscr{J}_3$ and $\mathscr{K}_3$ by μ and ν, respectively. These quantities have the following significance,

$$\mu = j_1 - j_2, \quad \nu = j_1 + j_2 + 1. \tag{3.6}$$

In evaluating the square of the resultant angular momentum, we encounter

$$2\,\mathbf{J}_1 \cdot \mathbf{J}_2 = \tfrac{1}{2} \sum_{\zeta\zeta'\zeta''\zeta'''} a_\zeta^+ a_{\zeta'} b_{\zeta''}^+ b_{\zeta'''}(\zeta|\boldsymbol{\sigma}|\zeta') \cdot (\zeta''|\boldsymbol{\sigma}|\zeta''')$$

$$= \sum_{\zeta\zeta'} a_\zeta^+ a_{\zeta'} b_{\zeta'}^+ b_\zeta - \tfrac{1}{2} n_1 n_2. \tag{3.7}$$

This can be expressed either in terms of the $\mathscr{J}$ operators, or of the $\mathscr{K}$ operators, since

$$\mathscr{J}_-\mathscr{J}_+ = \sum_{\zeta\zeta'} b_{\zeta'}^+ a_\zeta a_{\zeta'}^+ b_\zeta = n_2 + \sum_{\zeta\zeta'} a_\zeta^+ a_{\zeta'} b_{\zeta'}^+ b_\zeta, \tag{3.8}$$

and

$$\mathscr{K}_+\mathscr{K}_- = \sum_{\zeta\zeta'} a_\zeta^+ b_{\zeta'}^+ (a_{\zeta'} b_{\zeta''} - a_{\zeta'} b_\zeta) = n_1 n_2 - \sum_{\zeta\zeta'} a_\zeta^+ a_{\zeta'} b_{\zeta'}^+ b_\zeta. \tag{3.9}$$

Indeed,

$$\mathbf{J}^2 = \mathscr{J}_3(\mathscr{J}_3 + 1) + \mathscr{J}_-\mathscr{J}_+ = \mathscr{J}_3(\mathscr{J}_3 - 1) + \mathscr{J}_+\mathscr{J}_-, \tag{3.10}$$

and

$$\mathbf{J}^2 = \mathscr{K}_3(\mathscr{K}_3 - 1) - \mathscr{K}_+\mathscr{K}_- = \mathscr{K}_3(\mathscr{K}_3 + 1) - \mathscr{K}_-\mathscr{K}_+. \tag{3.11}$$

From the first, conventional, representation of $\mathbf{J}^2$ in terms of the angular momentum $\mathscr{J}$, we infer that

$$j \geq |\mu|, \tag{3.12}$$

or

$$j \geq |j_1 - j_2|, \tag{3.13}$$

while the hyperbolic representation implies that

$$\nu - 1 \geq j, \tag{3.14}$$

or

$$j_1 + j_2 \geq j. \tag{3.15}$$

We have thus arrived at

$$j_1 + j_2 \geq j \geq |j_1 - j_2|, \tag{3.16}$$

the familiar restriction on the composition of two angular momenta.

An eigenvector of $\mathbf{J}^2$ is conveniently labelled by the eigenvalues of J_3, $\mathscr{J}_3$, and $\mathscr{K}_3$. In virtue of (3.6), the resulting eigenvector $\Psi(jm\mu\nu)$ is equivalently designated as $\Psi(j_1 j_2 jm)$. In particular, the state with $\nu = j+1$ corresponds to $j_1 + j_2 = j$, and $2j_1 = j + \mu$, $2j_2 = j - \mu$. The special state of this type with $m=j$ can be realized in only one way, since $m=j_1+j_2$ requires that $m_1=j_1$, $m_2=j_2$. Thus

$$\Psi(jj\mu j+1) = \frac{(a_+^+)^{j+\mu}}{((j+\mu)!)^{1/2}} \frac{(b_+^+)^{j-\mu}}{((j-\mu)!)^{1/2}} \Psi_0. \tag{3.17}$$

With an arbitrary reference system, this result becomes

$$((2j)!)^{1/2} \sum_{m=-j}^{j} \phi_{jm}(x)\Psi(jm\mu j+1) = \frac{(xa^+)^{j+\mu}(xb^+)^{j-\mu}}{[(j+\mu)!\,(j-\mu)!]^{1/2}} \Psi_0, \tag{3.18}$$

according to (2.4). We multiply this $\mathscr{J}$ analogue of (1.13) with $\phi_{j\mu}(\xi)$, and sum with respect to μ,

$$((2j)!)^{1/2} \sum_{m\mu} \phi_{jm}(x)\phi_{j\mu}(\xi)\Psi(jm\mu j+1) = \frac{(\xi_+(xa^+)+\xi_-(xb^+))^{2j}}{(2j)!} \Psi_0. \tag{3.19}$$

Further summation with respect to j then yields

$$\sum_{jm\mu} ((2j)!)^{1/2}\phi_{jm}(x)\phi_{j\mu}(\xi)\Psi(jm\mu j+1) = e^{\xi_+(xa^+)+\xi_-(xb^+)} \Psi_0. \tag{3.20}$$

To complete the determination of the eigenvector $\Psi(jm\mu\nu)$, we need the analogue of (1.29), specifying the eigenvector with arbitrary ν in terms of that with the minimum value, $j+1$. For this purpose, we examine the operator[6]

$$V = t^{2\mathscr{K}_3 - 1} \tag{3.21}$$

which has the following significant properties,

$$t\frac{\partial}{\partial t}V = (2\mathscr{K}_3-1)V, \quad \left(t\frac{\partial}{\partial t}\right)^2 V = (2\mathscr{K}_3-1)^2 V, \tag{3.22}$$

and

$$V^{-1}\mathscr{K}_-V = t^2\mathscr{K}_-, \quad \mathscr{K}_-V = t^2 V\mathscr{K}_-. \tag{3.23}$$

In conjunction with

$$4\mathbf{J}^2 + 1 = (2\mathscr{K}_3-1)^2 - 4\mathscr{K}_+\mathscr{K}_-, \tag{3.24}$$

we obtain

$$\left(\frac{\partial^2}{\partial t^2} + \frac{1}{t}\frac{\partial}{\partial t} - \frac{4\mathbf{J}^2+1}{t^2}\right)V - 4\mathscr{K}_+V\mathscr{K}_- = 0, \tag{3.25}$$

an ordered operator form of Bessel's equation. The solution is

$$V = t^{(4\mathbf{J}^2+1)^{\frac{1}{2}}} F_{(4\mathbf{J}^2+1)^{\frac{1}{2}}}(t^2\mathscr{K}_+; P; \mathscr{K}_-), \tag{3.26}$$

where P is an integration constant, and the notation is intended to indicate that P is inserted between the powers of $\mathscr{K}_+$ and $\mathscr{K}_-$ in the ordered operator

---

[6] Our procedure here is based upon the general method of Appendix A.

expansion of the function F defined in (1.36). The second solution of the Bessel equation has been rejected in order to conform with the fact that $t^{2\mathcal{K}_3-1}$ must vanish as $t \to 0$, in view of the non-negative character of $\mathcal{K}_3 - 1$. The operator (3.26) can also be written as

$$V = \sum_j t^{2j+1} F_{2j+1}(t^2 \mathcal{K}_+; P_{j,j+1}; \mathcal{K}_-)$$

$$= \sum_j \sum_{\nu=j+1}^{\infty} t^{2\nu-1} P_{j\nu}, \tag{3.27}$$

where $P_{j\nu}$ is the projection operator for the state with the indicated eigenvalues. According to the well-known Bessel function power series we then have

$$P_{j\nu} = \omega_{j\nu}(\mathcal{K}_+) P_{j,j+1} \omega_{j\nu}(\mathcal{K}_-), \tag{3.28}$$

where

$$\omega_{j\nu}(\lambda) = \left[\frac{(2j+1)!}{(\nu+j)!\,(\nu-j-1)!}\right]^{1/2} \lambda^{\nu-j-1}. \tag{3.29}$$

This yields the desired eigenvector relation,

$$\Psi'(jm\mu\nu) = \omega_{j\nu}(\mathcal{K}_+) \Psi'(jm\mu j + 1). \tag{3.30}$$

It will be noted that, with respect to j and ν, Eq. (3.30) is converted into (1.29) by the substitutions

$$j \to |m| - \tfrac{1}{2}, \qquad \nu \to j + \tfrac{1}{2}, \tag{3.31}$$

which are in accord with the significance of K. Corresponding, then, to the generating functions (1.35) and (1.37), we have

$$\sum_{\nu=j+1}^{\infty} \omega_{j\nu}(\lambda) \Psi'(jm\mu\nu) = F_{2j+1}(\lambda \mathcal{K}_+) \Psi'(jm\mu j + 1), \tag{3.32}$$

and

$$((2j+1)!)^{-1/2} \sum_{\nu=j+1}^{\infty} \chi_{j\nu}(\lambda) \Psi'(jm\mu\nu) = e^{\lambda \mathcal{K}_+} \Psi'(jm\mu j + 1), \tag{3.33}$$

in which

$$\chi_{j\nu}(\lambda) = \left[\frac{(\nu+j)!}{(\nu-j-1)!}\right]^{1/2} \lambda^{\nu-j-1}. \tag{3.34}$$

The application of the operator $e^{\lambda \mathcal{K}_+}$ to (3.20) thus produces

$$\sum_{jm\mu\nu} (2j+1)^{-1/2} \phi_{jm}(x) \phi_{j\mu}(\xi) \chi_{j\nu}(\lambda) \Psi'(jm\mu\nu) = e^{\lambda[a+b^+] + \xi_+(xa^+) + \xi_-(xb^+)} \Psi'_0. \tag{3.35}$$

The eigenvectors are exhibited somewhat more explicitly[7] in the result

---

[7] The normalization constant does not automatically appear in the corresponding group theory formula. B. L. van der Waerden, "Die gruppentheoretische Methode in der Quantenmechanik" (Berlin, 1932).

obtained by applying $\omega_{j\nu}(\mathscr{K}_+)$ to (3.18),

$$\sum_{m=-j}^{j} \phi_{jm}(x)\Psi(j_1j_2jm) = \left[\frac{2j+1}{(j_1+j_2+j+1)!}\right]^{1/2}$$
$$\cdot \left[\frac{[a^+b^+]^{j_1+j_2-j}\cdot(xa^+)^{j+j_1-j_2}\cdot(xb^+)^{j_2+j-j_1}}{[(j_1+j_2-j)!(j+j_1-j_2)!(j_2+j-j_1)!]^{1/2}}\right]\Psi_0, \quad (3.36)$$

in which we have employed j_1 and j_2, rather than μ and ν. For the purpose of converting (3.36) into a convenient expression for the transformation function

$$(j_1j_2jm|j_1m_1j_2m_2) = (\Psi(j_1j_2jm), \Psi(j_1m_1j_2m_2)), \quad (3.37)$$

we make the replacement $x_+ \to z_-^*$, $x_- \to -z_+^*$, and take the scalar product with the generating function of the $\Psi(j_1m_1j_2m_2)$,

$$\sum_{j_1m_1j_2m_2} \phi_{j_1m_1}(x)\phi_{j_2m_2}(y)\Psi(j_1m_1j_2m_2) = e^{(xa^+)+(yb^+)}\Psi_0. \quad (3.38)$$

The ensuing formula can be written

$$\sum_{m_1m_2m_3} \phi_{j_1m_1}(x)\phi_{j_2m_2}(y)\phi_{j_3m_3}(z)X(j_1j_2j_3; m_1m_2m_3)$$
$$= [(j_1+j_2+j_3+1)!]^{-1/2} \cdot \frac{[yz]^{j_2+j_3-j_1}\cdot[zx]^{j_3+j_1-j_2}\cdot[xy]^{j_1+j_2-j_3}}{[(j_2+j_3-j_1)!(j_3+j_1-j_2)!(j_1+j_2-j_3)!]^{1/2}}, \quad (3.39)$$

in virtue of the definition[8]

$$(j_1j_2jm|j_1m_1j_2m_2) = (2j+1)^{1/2}(-1)^{j_1-j_2+m}X(j_1j_2j; m_1m_2-m). \quad (3.40)$$

Multiplication with

$$\Phi_{j_1j_2j_3}(\alpha\beta\gamma) = [(J+1)!]^{1/2}\frac{\alpha^{J-2j_1}\beta^{J-2j_2}\gamma^{J-2j_3}}{[(J-2j_1)!(J-2j_2)!(J-2j_3)!]^{1/2}}$$
$$J = j_1+j_2+j_3, \quad (3.41)$$

and summation with respect to j_1, j_2, and j_3, then yields the generating function

$$\sum_{jm} \phi_{j_1m_1}(x)\phi_{j_2m_2}(y)\phi_{j_3m_3}(z)\Phi_{j_1j_2j_3}(\alpha\beta\gamma)X(j_1j_2j_3; m_1m_2m_3) = e^{\alpha[yz]+\beta[zx]+\gamma[xy]}. \quad (3.42)$$

---

[8] This X coefficient is related to the V coefficient of G. Racah, *Phys. Rev.* **62**, 438 (1942), by $X = (-1)^{j_2+j-j_1}V$. We have introduced the X coefficient by virtue of its greater symmetry: compare Eqs. (3.44), (3.45) with Eq. (19a) of Racah's paper (henceforth referred to as R). Editors' note: This X coefficient is identical to the Wigner 3-j symbol; i.e., $X(j_1j_2j; m_1m_2m) = \begin{pmatrix} j_1 & j_2 & j \\ m_1 & m_2 & m \end{pmatrix}$.

Symmetry properties of the X coefficients can be easily inferred from the invariance of the generating function to particular substitutions. Thus, the null effect of multiplying x_+, y_+, z_+ by $e^{(i/2)\psi}$, and x_-, y_-, z_- by $e^{-(i/2)\psi}$, indicates that X vanishes unless

$$m_1 + m_2 + m_3 = 0. \tag{3.43}$$

The invariance of the generating function for simultaneous cyclic permutations of x, y, z and α, β, γ implies the corresponding property for X:

$$X(j_1 j_2 j_3; m_1 m_2 m_3) = X(j_2 j_3 j_1; m_2 m_3 m_1) = X(j_3 j_1 j_2; m_3 m_1 m_2). \tag{3.44}$$

The interchange of x and y, combined with the substitutions $\alpha \leftrightarrow -\beta$, $\gamma \to -\gamma$, discloses the behavior of the X coefficients with respect to non-cyclic permutations,

$$X(j_2 j_1 j_3; m_2 m_1 m_3) = X(j_1 j_3 j_2; m_1 m_3 m_2) = X(j_3 j_2 j_1; m_3 m_2 m_1)$$
$$= (-1)^J X(j_1 j_2 j_3; m_1 m_2 m_3), \tag{3.45}$$

while the exchange of x_+, y_+, z_+ with x_-, y_-, z_-, in conjunction with sign reversals for α, β, γ, leads to

$$X(j_1 j_2 j_3; -m_1 -m_2 -m_3) = (-1)^J X(j_1 j_2 j_3; m_1 m_2 m_3). \tag{3.46}$$

Among the implied properties of the transformation function (3.37) are

$$(j_2 j_1 jm | j_2 m_2 j_1 m_1) = (j_1 j_2 j - m | j_1 - m_1 j_2 - m_2)$$
$$= (-1)^{j_1 + j_2 - j}(j_1 j_2 jm | j_1 m_1 j_2 m_2). \tag{3.47}$$

The expression for $X(j_1 j_2 j_3; m_1 m_2 m_3)$, obtained by expanding (3.39), is

$$X(j; m) = [(J+1)!]^{-1/2} \sum_n (-1)^n \prod_{i=1}^3 \frac{[(j_i + m_i)!(j_i - m_i)!(J - 2j_i)!]^{1/2}}{(J - 2j_i - n_i)! n_i!} \tag{3.48}$$

in which

$$n = n_1 + n_2 + n_3, \tag{3.49}$$

and the summation is to be extended over all n_i subject to

$$J - 2j_i \geq n_i \geq 0, \tag{3.50}$$

and

$$n_2 - n_3 = m_1 - j_2 + j_3, \quad n_3 - n_1 = m_2 - j_3 + j_1, \quad n_1 - n_2 = m_3 - j_1 + j_2. \tag{3.51}$$

The latter conditions can also be written as

$$J - 2j_1 - n_1 = j_2 + m_2 - n_3 = j_3 - m_3 - n_2$$
$$J - 2j_2 - n_2 = j_3 + m_3 - n_1 = j_1 - m_1 - n_3$$
$$J - 2j_3 - n_3 = j_1 + m_1 - n_2 = j_2 - m_2 - n_1. \tag{3.52}$$

It follows from the non-negative character of these quantities that the n_i are uniquely determined if one of the nine integers $J - 2j_i, j_i + m_i, j_i - m_i$ is equal to zero. In general, the number of terms in the sum (3.48) exceeds by unity the smallest of these nine integers. It is a matter of convenience which of the n_i is chosen as the summation parameter.

The X coefficient can also be exhibited in closed form whenever the $|m_i|$ have the minimum values compatible with the given j_i. The simplest illustration of this is provided by $X(j_1 j_2 j_3; 000)$ corresponding to integral values of j_1, j_2, and j_3. Note that this quantity vanishes, according to (3.46), if $\tfrac{1}{2}J$ is not an integer. Our procedure here is to place $x_- = \partial/\partial x_+$, with analogous substitutions for y_- and z_-, and to evaluate the derivatives at $x_+ = y_+ = z_+ = 0$. Since

$$[(j_1 + m_1)!(j_1 - m_1)!]^{-1/2} \cdot (\partial/\partial x_+)^{j_1-m_1} x_+^{j_1+m_1}]_{x_+=0} = \delta_{m_1,0}, \quad (3.53)$$

this effectively isolates the $m = 0$ terms in (3.42). The reduction of the generating function can be performed with the aid of the following theorem concerning ordered operators, which will be proved in Appendix B. If a and a^+ are two operators satisfying $[a, a^+] = 1$, and $f(a^+)$ is an arbitrary function, we have

$$e^{za; a^+} f(a^+) = \frac{1}{1-z} f\left(\frac{a^+}{1-z}\right) e^{\frac{z}{1-z} a^+; a}. \quad (3.54)$$

The differential operator realization of this, with $a = \partial/\partial a^+$, is the form actually employed.

The result of the calculation is

$$\sum_j \Phi_{j_1 j_2 j_3}(\alpha \beta \gamma) X(j_1 j_2 j_3; 000) = (1 + \alpha^2 + \beta^2 + \gamma^2)^{-1}, \quad (3.55)$$

which is a generating function for $X(j; 0)$. On writing

$$(1 + \alpha^2 + \beta^2 + \gamma^2)^{-1} = \sum_{J=0, 2, \cdots} (-1)^{\tfrac{1}{2}J} (\alpha^2 + \beta^2 + \gamma^2)^{\tfrac{1}{2}J}$$

$$= \sum_j (-1)^{\tfrac{1}{2}J} (\tfrac{1}{2}J)! \frac{\alpha^{J-2j_1} \beta^{J-2j_2} \gamma^{J-2j_3}}{(\tfrac{1}{2}J - j_1)!(\tfrac{1}{2}J - j_2)!(\tfrac{1}{2}J - j_3)!}, \quad (3.56)$$

we obtain the explicit formula[9]

$$X(j; 0) = (-1)^{\tfrac{1}{2}J} \frac{(\tfrac{1}{2}J)!}{[(J+1)!]^{1/2}} \prod_{i=1}^{3} \frac{[(J-2j_i)!]^{1/2}}{(\tfrac{1}{2}J - j_i)!}. \quad (3.57)$$

We extend this argument by making the substitutions $x_- \to \partial/\partial x_+$, $y_- \to \partial/\partial y_+$, $z_+ \to \partial/\partial z_-$, and evaluating the derivatives for arbitrary x_+, y_+,

---

[9] This result is contained in R, Eq. (22′).

and z_-. In view of

$$\phi_{j_1 m_1}(x) \to \begin{cases} \left[\dfrac{(j_1+m_1)!}{(j_1-m_1)!}\right]^{1/2} \dfrac{x_+^{2m_1}}{(2m_1)!}, & m_1 \geq 0 \\ 0, & m_1 < 0, \end{cases} \tag{3.58}$$

and

$$\phi_{j_3 m_3}(z) \to \begin{cases} \left[\dfrac{(j_3+|m_3|)!}{(j_3-|m_3|)!}\right]^{1/2} \dfrac{z_-^{2|m_3|}}{(2|m_3|)!}, & m_3 \leq 0 \\ 0, & m_3 > 0, \end{cases} \tag{3.59}$$

we shall thereby obtain the X coefficient for $m_1 \geq 0, m_2 \geq 0, -m_3 = m_1 + m_2$. The values of X when two of the m_i are negative can then be inferred from (3.46). The generating function now becomes

$$e^{\alpha[yz]+\beta[zx]+\gamma[xy]} \to (1+\alpha^2+\beta^2+\gamma^2)^{-1}$$

$$\cdot \exp\left\{\dfrac{z_-}{1+\alpha^2+\beta^2+\gamma^2}\left[(\alpha\gamma-\beta)x_+ + (\beta\gamma-\alpha)y_+\right]\right\} \tag{3.60}$$

and, on expanding in powers of x_+, y_+, and z_-, we find that

$$\sum_j \Phi_{j_1 j_2 j_3}(\alpha\beta\gamma) \prod_i \left[\dfrac{(j_i+|m_i|)!}{(j_i-|m_i|)!}\right]^{1/2} X(j;m)$$

$$= (2|m_3|)! \dfrac{(\alpha\gamma-\beta)^{2m_1}(\beta\gamma-\alpha)^{2m_2}}{(1+\alpha^2+\beta^2+\gamma^2)^{2|m_3|+1}}. \tag{3.61}$$

The result attained by further expansion of (3.61) is

$$\left[(J+1)! \prod_i \dfrac{(j_i+|m_i|)!}{(j_i-|m_i|)!} \dfrac{1}{(J-2j_i)!}\right]^{1/2} X(j;m)$$

$$= \sum_{n_1 n_2} (-1)^{\frac{1}{2}J_2 - 2|m_3|} \dfrac{(\frac{1}{2}J_3)!}{\prod_i (\frac{1}{2}J_i - j_i - |m_i|)!} \dfrac{(2m_1)!}{(2m_1-n_1)! n_1!} \dfrac{(2m_2)!}{(2m_2-n_2)! n_2!} \tag{3.62}$$

where

$$J_1 = J + n_1 - n_2, \qquad J_2 = J - n_1 + n_2, \qquad J_3 = J + n_1 + n_2. \tag{3.63}$$

The double summation is to be extended over such non-negative integers that satisfy

$$J - 2j_1 - n_2 \geq 2m_1 - n_1 \geq 0$$
$$J - 2j_2 - n_1 \geq 2m_2 - n_2 \geq 0$$
$$J - 2j_3 \geq 2|m_3| - n_1 - n_2 \geq 0, \tag{3.64}$$

and for which $J + n_1 + n_2$ is an even integer. The sum consists of a single term if one of the $J - 2j_i$ vanishes, or if $m_1 = m_2 = 0$. This simplification may also result from the evenness requirement on J_3. Thus

$$\left[\frac{(j_1+\tfrac{1}{2})(j_3+\tfrac{1}{2})(J+1)!}{\prod_i(J-2j_i)!}\right]^{\tfrac{1}{2}} X(j_1j_2j_3; \tfrac{1}{2}0-\tfrac{1}{2})$$

$$= (-1)^{\tfrac{1}{2}J-1} \cdot \frac{(\tfrac{1}{2}J)!}{(\tfrac{1}{2}J-j_1-\tfrac{1}{2})!(\tfrac{1}{2}J-j_2)!(\tfrac{1}{2}J-j_3-\tfrac{1}{2})!}, \quad J \text{ even}$$

$$= (-1)^{\tfrac{1}{2}J+\tfrac{1}{2}} \cdot \frac{(\tfrac{1}{2}J+\tfrac{1}{2})!}{(\tfrac{1}{2}J-j_1)!(\tfrac{1}{2}J-j_2-\tfrac{1}{2})!(\tfrac{1}{2}J-j_3)!}, \quad J \text{ odd}, \quad (3.65)$$

which are the X coefficients with the minimum $|m_i|$ corresponding to half-integral values for two of the j_i.

The orthogonality and normalization of the eigenvectors $\Psi'(jm\mu\nu)$ can be verified, with the aid of (3.35), by an extension of the procedure leading to (1.21). According to Eq. (C7) of Appendix C, we have

$$(\exp\{\lambda[a^+b^+] + \xi_+(xa^+) + \xi_-(xb^+)\}\Psi_0,$$

$$\exp\{\kappa[a^+b^+] + \eta_+(ya^+) + \eta_-(yb^+)\}\Psi_0) = \frac{1}{(1-\lambda^*\kappa)^2}\exp\left[\frac{(\xi^*\eta)(x^*y)}{1-\lambda^*\kappa}\right],$$
(3.66)

and the expansion

$$\frac{1}{(1-\lambda^*\kappa)^2}\exp\frac{[(\xi^*\eta)(x^*y)]}{1-\lambda^*\kappa}$$

$$= \sum_{jm\mu\nu} \frac{1}{2j+1} \phi_{jm}(x^*)\phi_{jm}(y)\phi_{j\mu}(\xi^*)\phi_{j\mu}(\eta)\chi_{j\nu}(\lambda^*)\chi_{j\nu}(\kappa) \quad (3.67)$$

establishes that

$$(\Psi'(jm\mu\nu), \Psi'(j'm'\mu'\nu')) = \delta_{jj'}\,\delta_{mm'}\,\delta_{\mu\mu'}\,\delta_{\nu\nu'}. \quad (3.68)$$

The unitary nature of the transformation $\Psi'(j_1m_1j_2m_2) \to \Psi'(j_1j_2jm)$, and of its inverse, imposes the following conditions upon the X coefficients,

$$\sum_{m_1m_2} X(j_1j_2j_3; m_1m_2m_3)X(j_1j_2j_3'; m_1m_2m_3') = \frac{1}{2j_3+1}\delta_{j_3j_3'}\delta_{m_3m_3'} \quad (3.69)$$

and

$$\sum_{j_3m_3} (2j_3+1)X(j_1j_2j_3; m_1m_2m_3)X(j_1j_2j_3; m_1'm_2'm_3) = \delta_{m_1m_1'}\delta_{m_2m_2'}. \quad (3.70)$$

As a particular consequence of (3.69), we have

$$\sum_m [X(j; m)]^2 = 1. \quad (3.71)$$

The Rotation Matrices

The results of this section can be applied in developing further the properties of the matrices $U^{(j)}_{mm'}(\phi\theta\psi)$, which were introduced in Section 2. If U is the operator generating a reference system rotation for the composite system with angular momentum $\mathbf{J} = \mathbf{J}_1 + \mathbf{J}_2$, while U_1 and U_2 are the corresponding operators for the individual angular momenta, we have

$$U = U_1 U_2, \qquad (3.72)$$

according to the exponential form (2.31). In particular, theorem (2.59) states that

$$\int U_1 U_2 d\omega = P_0, \qquad (3.73)$$

where P_0 is the projection operator for the $j = 0$ state of the resultant angular momentum. On taking matrix elements of the latter equation, we find

$$\int U^{(j_1)}_{m_1 m'_1}(\omega) U^{(j_2)}_{m_2 m'_2}(\omega) d\omega = (j_1 m_1 j_2 m_2 | P_0 | j_1 m'_1 j_2 m'_2)$$

$$= (j_1 m_1 j_2 m_2 | j_1 j_2 00)(j_1 j_2 00 | j_1 m'_1 j_2 m'_2)$$

$$= \frac{1}{2j_1 + 1} \delta_{j_1 j_2} \delta_{-m_1 m_2} \delta_{-m'_1 m'_2}(-1)^{m_1 - m'_2} \qquad (3.74)$$

since

$$(j_1 j_2 00 | j_1 m_1 j_2 m_2) = (2j_1 + 1)^{-1/2}(-1)^{j_1 - m_1} \delta_{j_1 j_2} \delta_{-m_1 m_2}. \qquad (3.75)$$

In view of (2.41), it is also possible to write (3.74) as

$$\int U^{(j_1)}_{m_1 m'_1}{}^*(\omega) U^{(j_2)}_{m_2 m'_2}(\omega) d\omega = \frac{1}{2j_1 + 1} \delta_{j_1 j_2} \delta_{m_1 m_2} \delta_{m'_1 m'_2}, \qquad (3.76)$$

which expresses the orthogonality properties of the rotation matrices, in their dependence upon the rotation parameters.

The orthogonality relation of the trace $\chi^{(j)}$, derived from (3.76), is

$$\int \chi^{(j_1)*} \chi^{(j_2)} d\omega = \delta_{j_1 j_2}. \qquad (3.77)$$

This integral can be simplified, since the $\chi^{(j)}$ depend only upon the rotation angle γ. We write

$$d\omega = \int_0^{2\pi} \tfrac{1}{2} d\gamma \sin \tfrac{1}{2}\gamma \, \delta\!\left(\cos \tfrac{1}{2}\gamma - \cos \tfrac{1}{2}\theta \cdot \cos \frac{\phi + \psi}{2}\right) d\omega, \qquad (3.78)$$

and, after first performing the $d\omega$ integration, obtain

$$\int_0^{2\pi} \chi^{(j_1)}(\gamma)^* \chi^{(j_2)}(\gamma) \frac{1}{\pi} \sin^2 \frac{\gamma}{2} d\gamma = \delta_{j_1 j_2}, \qquad (3.79)$$

which can be verified directly.

We return to (3.72) and observe that its matrix element is

$$U^{(j_1)}_{m_1 m'_1}(\omega) U^{(j_2)}_{m_2 m'_2}(\omega) = \sum_{jmm'} (j_1 m_1 j_2 m_2 | j_1 j_2 j m) U^{(j)}_{mm'}(\omega)(j_1 j_2 j m | j_1 m'_1 j_2 m'_2)$$

$$= \sum_{jmm'} (2j+1) X(j_1 j_2 j; m_1 m_2 - m)(-1)^{m-m'} U^{(j)}_{mm'}(\omega) X(j_1 j_2 j; m'_1 m'_2 - m'), \quad (3.80)$$

or

$$U^{(j_1)}_{m_1 m'_1}(\omega) U^{(j_2)}_{m_2 m'_2}(\omega)$$
$$= \sum_{jmm'} (2j+1) X(j_1 j_2 j; m_1 m_2 m) U^{(j)}_{mm'}(\omega)^* X(j_1 j_2 j; m'_1 m'_2 m'). \quad (3.81)$$

With the use of the orthogonality relation (3.76), this can be presented in the symmetrical form

$$\int U^{(j_1)}_{m_1 m'_1} U^{(j_2)}_{m_2 m'_2} U^{(j_3)}_{m_3 m'_3} d\omega = X(j_1 j_2 j_3; m_1 m_2 m_3) X(j_1 j_2 j_3; m'_1 m'_2 m'_3). \quad (3.82)$$

Specializations of this integral are provided by

$$U^{(l)}_{m0} = \left(\frac{4\pi}{2l+1}\right)^{1/2} Y^*_{lm}(\theta\phi), \quad (3.83)$$

and

$$U^{(l)}_{00} = P_l(\cos\theta), \quad (3.84)$$

where Y_{lm} is the spherical harmonic associated with integral l, and $P_l(\cos\theta)$ is the Legendre polynomial.
Thus

$$\int Y_{l_1 m_1} Y_{l_2 m_2} Y_{l_3 m_3} \tfrac{1}{2}\sin\theta\, d\theta\, \frac{1}{2\pi} d\phi = \left[\prod_i \left(\frac{2l_i+1}{4\pi}\right)\right]^{1/2} X(l; 0)\, X(l; m), \quad (3.85)$$

and

$$\int_0^\pi P_{l_1}(\cos\theta) P_{l_2}(\cos\theta) P_{l_3}(\cos\theta) \tfrac{1}{2}\sin\theta\, d\theta = [X(l; 0)]^2. \quad (3.86)$$

The multiplication property of the trace, as derived from (3.80) is

$$\chi^{(j_1)}(\gamma)\chi^{(j_2)}(\gamma) = \sum_{j=|j_1-j_2|}^{j=j_1+j_2} \chi^{(j)}(\gamma), \quad (3.87)$$

which can also be expressed in the form

$$\int_0^{2\pi} \chi^{(j_1)}\chi^{(j_2)}\chi^{(j_3)} \frac{1}{\pi}\sin^2 \tfrac{1}{2}\gamma\, d\gamma = \begin{cases} 1, & J-2j_i \geq 0 \\ 0, & \text{otherwise.} \end{cases} \quad (3.88)$$

One can regard this as a realization of the projection operator statement of the angular momentum composition law,

$$P_{j_1}P_{j_2} = \sum_{j=|j_1-j_2|}^{j=j_1+j_2} P_j, \qquad (3.89)$$

since (3.87) is the trace of the equation obtained by multiplying (3.89) with $U_1 U_2 = U$.

We shall conclude this discussion by deriving the completeness relations for the functions $\chi^{(j)}(\omega)$ and $U^{(j)}_{mm'}(\omega)$. Referring to (2.36), the generating function of the $\chi^{(j)}$, we replace t therein with $te^{(i/2)\gamma'}$ and obtain

$$\sum_j t^{2j} \chi^{(j)}(\omega) e^{i(j+\frac{1}{2})\gamma'} = \frac{1}{(1+t^2)\cos\frac{\gamma'}{2} - 2t\cos\frac{\gamma}{2} - i\sin\frac{\gamma'}{2}(1-t^2)}, \qquad (3.90)$$

the imaginary part of which can be written

$$\sum_j t^{2j}\chi^{(j)}(\omega)\chi^{(j)}(\omega')$$

$$= \frac{(1-t^2)}{(1-t^2)^2\left(1 - \frac{4t}{(1+t)^2}\cos\frac{\gamma}{2}\cos\frac{\gamma'}{2}\right) + 4t^2\left(\cos\frac{\gamma}{2} - \cos\frac{\gamma'}{2}\right)^2}. \qquad (3.91)$$

We now consider the limit $t \to 1$, and infer from the known result

$$\lim_{\varepsilon \to 0} \frac{1}{\pi} \frac{\varepsilon}{x^2 + \varepsilon^2} = \delta(x), \qquad (3.92)$$

that

$$\sum_j \chi^{(j)}(\omega)\chi^{(j)}(\omega') = \frac{\pi}{2} \frac{1}{\sin(\gamma/2)} \delta\left(\cos\frac{\gamma}{2} - \cos\frac{\gamma'}{2}\right). \qquad (3.93)$$

However

$$\int \frac{\pi}{2} \frac{1}{\sin\gamma/2} \delta\left(\cos\frac{\gamma}{2} - \cos\frac{\gamma'}{2}\right) d\omega = \int_0^{2\pi} \delta\left(\cos\frac{\gamma}{2} - \cos\frac{\gamma'}{2}\right)\sin\frac{\gamma}{2} \tfrac{1}{2} d\gamma = 1, \qquad (3.94)$$

so that (3.93) can be written

$$\sum_j \chi^{(j)}(\omega)\chi^{(j)}(\omega') = \delta(\omega - \omega'), \qquad (3.95)$$

which is the completeness relation of the $\chi^{(j)}$. As a specialization of (3.95), we place $\gamma' = 0$ and find

$$\sum_j (2j+1)\chi^{(j)}(\omega) = \delta(\omega). \qquad (3.96)$$

An operator expression for the composition of successive rotations is given by
$$U(\omega)U^{-1}(\omega') = U(\omega - \omega'). \tag{3.97}$$

We take the trace of this equation for the states with quantum number j, and, in virtue of the unitary property of U, obtain
$$\sum_{mm'} U^{(j)}_{mm'}(\omega) U^{(j)}_{mm'}(\omega')^* = \chi^{(j)}(\omega - \omega'), \tag{3.98}$$

which is in the nature of an addition theorem. The completeness relation for the $U^{(j)}_{mm'}(\omega)$ is reached on multiplying (3.98) with $2j+1$ and summing with respect to j. In view of (3.96), we have
$$\sum_{jmm'} (2j+1) U^{(j)}_{mm'}(\omega) U^{(j)}_{mm'}(\omega')^* = \delta(\omega - \omega'). \tag{3.99}$$

On integration of (3.98) and (3.99) with respect to the Eulerian angle ψ, there emerges the addition theorem and the completeness relation of the spherical harmonics.

4. Three and Four Angular Momenta

Eigenvectors for the resultant of three angular momenta can be built up in several ways, as symbolized by
$$\mathbf{J} = \mathbf{J}_1 + (\mathbf{J}_2 + \mathbf{J}_3) = \mathbf{J}_2 + (\mathbf{J}_3 + \mathbf{J}_1) = \mathbf{J}_3 + (\mathbf{J}_1 + \mathbf{J}_2). \tag{4.1}$$

Thus, according to the first procedure, we construct $\Psi(j_1 m_1 j_2 j_3 j_{23} m_{23})$ and then $\Psi(j_1[j_2 j_3] j_{23} jm)$, while the last method yields $\Psi(j_3[j_1 j_2] j_{12} jm)$. The notation $[j_2 j_3]$, for example, is intended to indicate that these angular momenta are not involved explicitly in the composition of j_1 and j_{23} to form j. Similarly, four angular momenta can be combined in various pairs,
$$\mathbf{J} = (\mathbf{J}_1 + \mathbf{J}_2) + (\mathbf{J}_3 + \mathbf{J}_4) = (\mathbf{J}_2 + \mathbf{J}_3) + (\mathbf{J}_4 + \mathbf{J}_1) = (\mathbf{J}_1 + \mathbf{J}_3) + (\mathbf{J}_2 + \mathbf{J}_4), \tag{4.2}$$

in which the first method, say, yields $\Psi([j_1 j_2] j_{12} [j_3 j_4] j_{34} jm)$ through the intermediary of $\Psi(j_1 j_2 j_{12} m_{12} j_3 j_4 j_{34} m_{34})$. Our problem in this section is the evaluation of the transformation function connecting two such schemes of adding four angular momenta. The analogous question for three angular momenta can be regarded as a specialization of this more symmetrical problem.

To facilitate the addition of angular momenta in pairs, we observe that the generating function (3.35), written as
$$\sum_{j_1 j_2 jm} (2j+1)^{-1/2} \phi_{jm}(x) \Phi_{j_1 j_2 j}(\alpha_1 \alpha_2 \alpha_3) \Psi(j_1 j_2 jm)$$
$$= \exp(\alpha_3[a^+ b^+] + \alpha_2(xa^+) + \alpha_1(xb^+)) \Psi_0, \tag{4.3}$$

can be obtained from

$$\sum_{j_1 m_1 j_2 m_2} \phi_{j_1 m_1}(t_1)\phi_{j_2 m_2}(t_2)\Psi'(j_1 m_1 j_2 m_2) = \exp((t_1 a^+) + (t_2 b^+))\Psi'_0 \quad (4.4)$$

by the application of the differential operator

$$\exp\left(\alpha_3\left[\frac{\partial}{\partial t_1}\frac{\partial}{\partial t_2}\right] + \alpha_2\left(x\frac{\partial}{\partial t_1}\right) + \alpha_1\left(x\frac{\partial}{\partial t_2}\right)\right) \quad (4.5)$$

with the understanding that the derivatives are to be evaluated at $t_1 = t_2 = 0$. Accordingly, if we apply (4.5) and

$$\exp\left(\beta_3\left[\frac{\partial}{\partial t_3}\frac{\partial}{\partial t_4}\right] + \beta_2\left(y\frac{\partial}{\partial t_3}\right) + \beta_1\left(y\frac{\partial}{\partial t_1}\right)\right), \quad (4.6)$$

to the generating function of the $\Psi'(j_1 m_1 j_2 m_2 j_3 m_3 j_4 m_4)$, namely,

$$\exp((t_1 a^+) + (t_2 b^+) + (t_3 c^+) + (t_4 d^+))\Psi'_0, \quad (4.7)$$

we shall obtain a function generating $\Psi'(j_1 j_2 j_{12} m_{12} j_3 j_4 j_{34} m_{34})$. The further application of the operator

$$\exp\left(\gamma_3\left[\frac{\partial}{\partial x}\frac{\partial}{\partial y}\right] + \gamma_2\left(z\frac{\partial}{\partial x}\right) + \gamma_1\left(z\frac{\partial}{\partial y}\right)\right) \quad (4.8)$$

then produces

$$\sum_{j_1 j_2 j_3 j_4 j_{12} j_{34} jm} [(2j_{12}+1)(2j_{34}+1)(2j+1)]^{-1/2}\phi_{jm}(z)\Phi_{j_1 j_2 j_{12}}(\alpha)$$
$$\cdot \Phi_{j_3 j_4 j_{34}}(\beta)\Phi_{j_{12} j_{34} j}(\gamma)\Psi'([j_1 j_2]j_{12}[j_3 j_4]j_{34} jm) = e^Q \Psi'_0, \quad (4.9)$$

in which

$$Q = \alpha_3[a^+ b^+] + \beta_3[c^+ d^+] + \gamma_3\alpha_1\beta_1[b^+ d^+] + \gamma_3\alpha_1\beta_2[b^+ c^+] + \gamma_3\alpha_2\beta_1[a^+ d^+]$$
$$+ \gamma_3\alpha_2\beta_2[a^+ c^+] + \gamma_2\alpha_2(za^+) + \gamma_2\alpha_1(zb^+) + \gamma_1\beta_2(zc^+) + \gamma_1\beta_1(zd^+). \quad (4.10)$$

As an important specialization of (4.9), yielding the eigenvectors with $j = 0$, we place $\gamma_1 = \gamma_2 = 0$, and $\gamma_3 = 1$, with the result

$$\sum_{j_1 j_2 j_3 j_4} (2j'+1)^{-1/2}\Phi_{j_1 j_2 j'}(\alpha)\Phi_{j_3 j_4 j'}(\beta)\Psi'([j_1 j_2]j'[j_3 j_4]j'00) = e^R \Psi'_0,$$

$$R = \alpha_3[a^+ b^+] + \beta_3[c^+ d^+] + \alpha_1\beta_1[b^+ d^+] + \alpha_1\beta_2[b^+ c^+] + \alpha_2\beta_1[a^+ d^+]$$
$$+ \alpha_2\beta_2[a^+ c^+], \quad (4.11)$$

where $j' = j_{12} = j_{34}$. An analogous equation for a different mode of addition is

$$\sum_{j_1 j_2 j_3 j_4} (2j''+1)^{-1/2}\Phi_{j_1 j_3 j''}(\alpha')\Phi_{j_2 j_4 j''}(\beta')\Psi'([j_1 j_3]j''[j_2 j_4]j''00) = e^{R'}\Psi'_0,$$

$$R' = \alpha'_3[a^+ c^+] + \beta'_3[b^+ d^+] + \alpha'_1\beta'_1[c^+ d^+] + \alpha'_1\beta'_2[c^+ b^+] + \alpha'_2\beta'_1[a^+ d^+]$$
$$+ \alpha'_2\beta'_2[a^+ b^+]. \quad (4.12)$$

The transformation function connecting the two schemes is determined by[10]

$$\sum_{j_1j_2j_3j_4j'j''} (-1)^{j'+j''-j_1-j_4} \Phi_{j_1j_2j'}(\alpha)\Phi_{j_3j_4j'}(\beta)\Phi_{j_1j_3j''}(\alpha')\Phi_{j_2j_4j''}(\beta')W(j_1j_2j_3j_4;j'j'')$$
$$= (e^R \Psi_0, e^R \Psi_0), \quad (4.13)$$

in which we have written[11]

$$([j_1j_2]j'[j_3j_4]j'0|[j_1j_3]j''[j_2j_4]j''0) = (-1)^{j'+j''-j_1-j_4}$$
$$\cdot [(2j'+1)(2j''+1)]^{1/2} W(j_1j_2j_3j_4;j'j''). \quad (4.14)$$

We now employ the theorem [Eq. (C28)]

$$(\exp\{\tfrac{1}{2}\sum_{\mu,\nu=1}^{4} \lambda_{\mu\nu}^{*}[A_\mu^+ A_\nu^+]\}\Psi_0, \; \exp\{\tfrac{1}{2}\sum_{\mu,\nu=1}^{4} \kappa_{\mu\nu}[A_\mu^+ A_\nu^+]\}\Psi_0)$$
$$= [1 - \tfrac{1}{2}\sum \lambda_{\mu\nu}^* \kappa_{\mu\nu} + |\lambda^*|^{1/2}|\kappa|^{1/2}]^{-2}, \quad (4.15)$$

in which the $A_{\zeta\mu}$ are four sets of two component operators, obeying

$$[A_{\zeta\mu}, A_{\zeta'\nu}^+] = \delta_{\mu\nu}\delta_{\zeta\zeta'}, \quad (4.16)$$

and $|\lambda|$, $|\kappa|$ are the determinants of the antisymmetrical matrices $\lambda_{\mu\nu}$ and $\kappa_{\mu\nu}$. For the application in question,

$$|\lambda|^{1/2}|\kappa|^{1/2} = -\alpha_3\beta_3\alpha_3'\beta_3',$$
$$\tfrac{1}{2}\sum_{\mu\nu} \lambda_{\mu\nu}\kappa_{\mu\nu} = \alpha_3\alpha_2'\beta_2' + \beta_3\alpha_1'\beta_1' + \alpha_3'\alpha_2\beta_2 + \beta_3'\alpha_1\beta_1 - \alpha_1\beta_2\alpha_1'\beta_2' + \alpha_2\beta_1\alpha_2'\beta_1'. \quad (4.17)$$

On changing the signs of α_1 and β_3', we obtain for the generating function of the W coefficients,

$$\sum_{j_1j_2j_3j_4j'j''} \Phi_{j_1j_2j'}(\alpha)\Phi_{j_3j_4j'}(\beta)\Phi_{j_1j_3j''}(\alpha')\Phi_{j_2j_4j''}(\beta')W(j_1j_2j_3j_4;j'j'')$$
$$= [1 - \alpha_3\alpha_2'\beta_2' - \beta_3\alpha_1'\beta_1' - \alpha_3'\alpha_2\beta_2 - \beta_3'\alpha_1\beta_1 - \alpha_1\beta_2\alpha_1'\beta_2' - \alpha_2\beta_1\alpha_2'\beta_1' + \alpha_3\beta_3\alpha_3'\beta_3']^{-2}. \quad (4.18)$$

The symmetry properties expressed by

$$W(j_1j_2j_3j_4;j'j'') = W(j_2j_1j_4j_3;j'j'') = W(j_3j_4j_1j_2;j'j'')$$
$$= W(j_1j_3j_2j_4;j''j') \quad (4.19)$$

follow from the invariance of (4.18) under the respective substitutions: $\alpha_1 \leftrightarrow \alpha_2$, $\beta_1 \leftrightarrow \beta_2$, $\alpha' \leftrightarrow \beta'$; $\alpha_1' \leftrightarrow \alpha_2'$, $\beta_1' \leftrightarrow \beta_2'$, $\alpha \leftrightarrow \beta$; $\alpha \leftrightarrow \alpha'$, $\beta \leftrightarrow \beta'$, while the more complicated transformation $(\alpha_1\alpha_2\alpha_3) \to (-\alpha_3\alpha_2\alpha_1)$, $(\alpha_1'\alpha_2'\alpha_3') \leftrightarrow (\beta_3\beta_1\beta_2)$, $(\beta_1'\beta_2'\beta_3') \to (\beta_1'\beta_3' - \beta_2')$ yields

$$W(j_1j_2j_3j_4;j'j'') = (-1)^{j'+j''-j_1-j_4} W(j'j_2j_3j'';j_1j_4). \quad (4.20)$$

---

[10] For simplicity we have assumed that the parameters α, β are real. The generating function (4.18) is valid without this restriction.

[11] The W coefficient thereby defined is the same as that discussed in R.

Twenty-four equivalent forms for W are obtained by repeated use of (4.19) and (4.20).

Further characteristics of W follow from the composition properties of the transformation function (4.14), which we shall temporarily indicate by $(12, 34j'|13, 24j'')$. Thus

$$\sum_{j''} (12, 34j'|13, 24j'')(13, 24j''|12, 34j''') = \delta_{j'j'''} \qquad (4.21)$$

and

$$\sum_{j''} (12, 34j'|13, 24j'')(13, 24j''|14, 23j''') = (12, 34j'|14, 23j'''). \qquad (4.22)$$

All of these quantities can be expressed in terms of W. The interchange of 2 and 4, and of 3 and 4 in (4.14) yields, with the aid of (3.47),

$$(13, 24j''|14, 23j''') = (-1)^{j_2+j_3+j_4-j_1}[(2j''+1)(2j'''+1)]^{1/2}W(j_1j_4j_3j_2;j'''j''), \qquad (4.23)$$

and

$$(12, 34j'|14, 23j''') = (-1)^{j'''+j_4-j_1}[(2j'+1)(2j'''+1)]^{1/2}W(j_1j_2j_4j_3;j'j'''). \qquad (4.24)$$

Therefore

$$\sum_{j''} (2j''+1)W(j_1j_2j_3j_4;j'j'')W(j_1j_2j_3j_4;j'''j'') = \frac{1}{2j'+1}\delta_{j'j'''} \qquad (4.25)$$

and

$$\sum_{j''} (-1)^{j'+j''+j'''+j_1+j_2+j_3+j_4}(2j''+1)W(j_1j_2j_3j_4;j'j'')W(j_1j_4j_3j_2;j'''j'')$$
$$= W(j_1j_2j_4j_3;j'j'''). \qquad (4.26)$$

These formulae can be combined by placing $j_2 = j_4$, $j' = j'''$ in (4.26) and, after multiplication with $2j'+1$, performing the summation with respect to j' by means of (4.25). We obtain

$$\sum_{j'} (2j'+1)W(j_1j_2j_2j_3;j'j') = \sum_{j''} (-1)^{j''-j_2+j_3}, \qquad (4.27)$$

in which the values assumed by j'' are those compatible with the existence of $W(j_1j_2j_3j_2;j'j'')$, namely, $j'' \geq |j_1-j_3|$, $j'' \leq j_1+j_3$, $2j_2$. Accordingly,

$$\sum_{j'} (2j'+1)W(j_1j_2j_2j_3;j'j') = \begin{cases} 1, & k \text{ even} \\ 0, & k \text{ odd} \end{cases} \qquad (4.28)$$

where k is the smaller of the two integers $j_1+j_3-|j_1-j_3|$, $2j_2-|j_1-j_3|$. One of the consequences of (4.28),

$$W(j_1j_2j_20;j_2j_2) = \frac{1}{2j_2+1}, \qquad j_1 \leq 2j_2, \qquad (4.29)$$

is a particular example of

$$W(j_1j_2j_30;j_3j_2)=[(2j_2+1)(2j_3+1)]^{-1/2}, |j_2-j_3|\leq j_1\leq j_2+j_3, \quad (4.30)$$

which follows from (4.14) on remarking that, with $j_4=0$, the interchange of j_2 and j_3 simply multiplies the eigenvector with $(-1)^{j_2+j_3-j_1}$.

The relation between the W and X coefficients can be inferred from (4.14) by writing

$$\Psi'([j_1j_2]j'[j_3j_4]j'0)$$

$$=\sum_{m_{12}}(2j'+1)^{-1/2}(-1)^{j'-m_{12}}\Psi(j_1j_2j'm_{12}j_3j_4j'-m_{12})$$

$$=(2j'+1)^{1/2}(-1)^{j_1+j_3-j_2-j_4}\sum_{m_{12}}X(j_1j_2j';m_1m_2-m_{12})(-1)^{j'-m_{12}}$$

$$\cdot X(j_3j_4j';m_3m_4m_{12})\Psi(j_1m_1j_2m_2j_3m_3j_4m_4), \quad (4.31)$$

which, with the similar representation of $\Psi([j_1j_3]j''[j_2j_4]j''0)$, yields

$$\sum_m X(j_1j_2j';m_1m_2-m_{12})(-1)^{j'-\dot{m}_{12}}X(j_3j_4j';m_3m_4m_{12})$$

$$\cdot X(j_1j_3j'';m_1m_3-m_{13})(-1)^{j''-m_{13}}X(j_2j_4j'';m_2m_4m_{13})$$

$$=(-1)^{j'+j''+j_1+j_4}W(j_1j_2j_3j_4;j'j''). \quad (4.32)$$

The general expression obtained for W by expanding the generating function (4.18) can be cast into the form

$$W(j_1j_2j_3j_4;j'j'')=\prod_r[(n+p+1-n_r)!]^{-1/2}\prod_{r,s}[(n_r+p_s)!]^{1/2}$$

$$\cdot \sum_{r,s}(-1)^{p_3}\frac{(n+p+1)!}{\prod n_r!p_s!}, \quad (4.33)$$

where

$$n=\sum_{r=1}^{4}n_r, \quad p=\sum_{s=1}^{3}p_s, \quad (4.34)$$

and the summation is to be extended over the non-negative integers, n_r, p_s, for which

$$j_2+j'-j_1-p_1=j_4+j'-j_3-p_2=j_2+j_4-j''-p_3=n_1,$$

$$j_3+j'-j_4-p_1=j_1+j'-j_2-p_2=j_1+j_3-j''-p_3=n_2,$$

$$j_3+j''-j_1-p_1=j_4+j''-j_2-p_2=j_3+j_4-j'-p_3=n_3,$$

$$j_2+j''-j_4-p_1=j_1+j''-j_3-p_2=j_1+j_2-j'-p_3=n_4. \quad (4.35)$$

The number of terms in the sum exceeds by unity the smallest of the twelve quantum number combinations, $j_2+j'-j_1$, etc.; the sum reduces to a single term if one such combination vanishes. The choice of summation parameter is a matter of convenience.

We now return to the general problem, that of evaluating the transformation function

$$([j_1j_2]j_{12}[j_3j_4]j_{34}jm|[j_1j_3]j_{13}[j_2j_4]j_{24}jm) \equiv (-1)^{j_{12}+j_{24}-j_1-j_4}$$
$$\cdot [(2j_{12}+1)(2j_{34}+1)(2j_{13}+1)(2j_{24}+1)]^{1/2} S(j_1j_2j_3j_4; j_{12}j_{34}j_{13}j_{24}; j). \quad (4.36)$$

A generating function for the S coefficient is given by[12]

$$\sum \Phi_{j_1j_2j_{12}}(\alpha) \Phi_{j_3j_4j_{34}}(\beta) \Phi_{j_{12}j_{34}j}(\gamma) \Phi_{j_1j_3j_{13}}(\alpha')$$
$$\cdot \Phi_{j_2j_4j_{24}}(\beta') \Phi_{j_{13}j_{24}j}(\gamma') S(j_1j_2j_3j_4; j_{12}j_{34}j_{13}j_{24}; j)$$
$$= [1 + \alpha_3\beta_3\alpha'_3\beta'_3 - \gamma_3(\alpha'_3\alpha_2\beta_2 + \beta'_3\alpha_1\beta_1) - \gamma'_3(\alpha_3\alpha'_2\beta'_2 + \beta_3\alpha'_1\beta'_1)$$
$$- \gamma_3\gamma'_3(\alpha_2\beta_1\alpha'_2\beta'_1 + \alpha_1\beta_2\alpha'_1\beta'_2) - \gamma_2\gamma'_2(\alpha_2\alpha'_2 + \beta_3\beta'_3\alpha_1\alpha'_1) - \gamma_1\gamma'_1(\beta_1\beta'_1 + \alpha_3\alpha'_3\beta_2\beta'_2)$$
$$- \gamma_1\gamma'_2(\beta_2\alpha'_1 + \alpha_3\beta'_3\beta_1\alpha'_2) + \gamma_2\gamma'_1(\alpha_1\beta'_2 + \alpha'_3\beta_3\alpha_2\beta'_1)]^{-2}, \quad (4.37)$$

where the sum is over all j's. The connection with the X coefficients is contained in

$$\sum X(j_1j_2j_{12}; m_1m_2-m_{12})X(j_3j_4j_{34}; m_3m_4-m_{34})X(j_{12}j_{34}j; m_{12}m_{34}-m)$$
$$\cdot X(j_1j_3j_{13}; m_1m_3-m_{13})X(j_2j_4j_{24}; m_2m_4-m_{24})X(j_{13}j_{24}j; m_{13}m_{24}-m)$$
$$= (-1)^{j_{34}+j_{13}+j_1+j_4+2j} S(j_1j_2j_3j_4; j_{12}j_{34}j_{13}j_{24}; j) \quad (4.38)$$

(the sum is over all m's) and the W coefficient appears as a special example,

$$S(j_1j_2j_3j_4; j'j''j'j''; 0) = [(2j'+1)(2j''+1)]^{-1/2} W(j_1j_2j_3j_4; j'j''). \quad (4.39)$$

In view of the complexity of the S coefficient we shall be content to record here only those cases that can be expressed in terms of W. This occurs whenever one of the nine quantum numbers involved in the S coefficient equals zero, which is a consequence of (4.39) and the fact that the symmetry of S is such that any of the other quantum numbers can appear in the position of j. Thus, it follows from either (4.37) or (4.38) that

$$S(j_1j_2j_3j_4; j_{12}j_{34}j_{13}j_{24}; j)$$
$$= (-1)^{j_{13}+j_{24}-j_1+j_2-j_3-j_4} S(j_{12}j_1j_{34}j_3; j_2j_4jj_{13}; j_{24})$$
$$= (-1)^{j_{24}+j_{34}-j_{12}-j_2-j_4-j} S(jj_{12}j_{13}j_1; j_{34}j_3j_{24}j_2; j_4), \quad (4.40)$$

which are representative of the eight permutations of this type. We obtain from (4.39) that

$$S(j_1j_2j_3j_2; j_{12}j_{34}j_{13}0; j_{13})$$
$$= (-1)^{j_{13}-j_1-j_3}[(2j_2+1)(2j_{13}+1)]^{-1/2} W(j_1j_{12}j_3j_{34}; j_2j_{13}), \quad (4.41)$$

---

[12] This is obtained with the aid of Eq. (C30).

and

$$S(j_1j_2j_30; j_{12}j_3j_{13}j_2; j)$$
$$= (-1)^{j_3-j_{12}-j}[(2j_2+1)(2j_3+1)]^{-1/2}W(jj_{12}j_{13}j_1; j_3j_2). \quad (4.42)$$

The latter result contains the solution to the problem of three angular momenta. Expressed in terms of a transformation function, without explicit reference to the angular momentum with zero quantum number, (4.42) states that

$$([j_1j_2]j_{12}j_3jm|[j_1j_3]j_{13}j_2jm)$$
$$= (-1)^{j_{12}+j_{13}-j_1-j}[(2j_{12}+1)(2j_{13}+1)]^{1/2}W(j_1j_2j_3j; j_{12}j_{13}). \quad (4.43)$$

A slightly simpler form[13] is obtained on permuting the indices 1 and 2, together with a change in sense of addition for j_1 and j_{23},

$$([j_1j_2]j_{12}j_3jm|j_1[j_2j_3]j_{23}jm)$$
$$= [(2j_{12}+1)(2j_{23}+1)]^{1/2}W(j_1j_2jj_3; j_{12}j_{23}). \quad (4.44)$$

As a particular consequence of this result, note that, according to (4.30),

$$([j_1j_2]j_3j_30|j_1[j_2j_3]j_10) = 1, \quad (4.45)$$

that is, the eigenvector for the null resultant of three angular momenta is independent of the mode of addition, provided that the order of the angular momenta is preserved. As one representation of this eigenvector we have

$$\Psi(j_1j_2j_30) = \sum_{m_3}[(2j_3+1)]^{-1/2}(-1)^{j_3+m_3}\Psi([j_1j_2]j_3-m_3j_3m_3), \quad (4.46)$$

and therefore

$$(j_1m_1j_2m_2j_3m_3|j_1j_2j_30) = [(2j_3+1)]^{-1/2}(-1)^{j_3+m_3}(j_1m_1j_2m_2|j_1j_2j_3-m_3)$$
$$= (-1)^{j_1+j_3-j_2}X(j_1j_2j_3; m_1m_2m_3), \quad (4.47)$$

in virtue of (3.40). Thus, the X coefficient, originally defined in terms of the addition of two angular momenta, now appears as characterizing three angular momenta with a null resultant.

This possibility, of replacing $\mathbf{J}_1 + \mathbf{J}_2 = \mathbf{J}$ with $\mathbf{J}_1 + \mathbf{J}_2 + \mathbf{J}_3 = 0$, depends upon the circumstance that the negative of an angular momentum operator is, in a certain sense, also an angular momentum operator. The commutation relations

$$\mathbf{J} \times \mathbf{J} = i\mathbf{J} \quad (4.48)$$

imply that

$$(-\mathbf{J}) \times (-\mathbf{J}) = -i(-\mathbf{J}), \quad (4.49)$$

---

[13] G. Racah, *Phys. Rev.* **63**, 367(1943).

which reassume the form (4.48) on changing the sign of i (complex, not Hermitian conjugation). Therefore

$$\mathbf{J}' = -\mathbf{J}^* \tag{4.50}$$

is an angular momentum operator. To find the eigenvectors of $\mathbf{J}'$, we notice that a rotation operator U is a function of $i\mathbf{J}$ and real angles. Therefore

$$U' = U^* \tag{4.51}$$

is the same function of $\mathbf{J}'$ that U is of $\mathbf{J}$. On taking the complex conjugate of the equation

$$U\Psi(jm') = \sum_m \Psi(jm) U^{(j)}_{mm'} \tag{4.52}$$

we obtain

$$U'\Psi^*(jm') = \sum_m \Psi^*(jm)(-1)^{m-m'} U^{(j)}_{-m,-m'}, \tag{4.53}$$

with the aid of (2.41). Hence

$$\Psi'(jm) = (-1)^{j+m}\Psi^*(j-m) \tag{4.54}$$

are the eigenvectors associated with $\mathbf{J}'$.

Now observe that the following dyadic, formed from the eigenvectors of a single angular momentum,

$$(2j+1)^{-1/2}\sum_m \Psi(jm)\Psi^*(jm), \tag{4.55}$$

is unchanged by a rotation of the reference system, since

$$\sum_{mm'm''} \Psi(jm')(jm'|U|jm)(jm|U^{-1}|jm'')\Psi^*(jm'') = \sum_m \Psi(jm)\Psi^*(jm). \tag{4.56}$$

Therefore, on employing (4.54) we infer that the *vector*

$$(2j+1)^{-1/2} \sum_m \Psi(jm)(-1)^{j-m}\Psi'(j-m) \tag{4.57}$$

describes the spherically symmetrical state of two angular momenta, which is in agreement with (3.75). This is the basic example of the relationship involved in (4.47).

5. Tensor Operators

An irreducible tensor operator of rank $j(=0, \frac{1}{2}, 1, ...)$ is a set of $2j+1$ operators, $T(jm)$, which transforms in the following manner under a change in coordinate system,

$$UT(jm')U^{-1} = \sum_{m=-j}^{j} T(jm) U^{(j)}_{mm'}. \tag{5.1}$$

On taking the Hermitian conjugate of this equation and employing (2.41), we find that $i^{2m}T(j-m)^\dagger$ transforms in the same manner as $T(jm)$. We therefore define the Hermitian conjugate tensor $T^\dagger$ according to

$$T^\dagger(jm) = i^{2m}(Tj-m)^\dagger. \tag{5.2}$$

The tensor that is conjugate to $T^\dagger$ is then described by

$$T^{\dagger\dagger}(jm) = i^{2m}(T^\dagger(j-m))^\dagger = i^{2m}(i^{-2m}T(jm)^\dagger)^\dagger = (-1)^{2m}T(jm), \tag{5.3}$$

or

$$T^{\dagger\dagger} = (-1)^{2j}T. \tag{5.4}$$

This shows that Hermitian tensors, $T\dagger = T$, exist only for integral j,[14] and satisfy

$$T(jm) = (-1)^m T(j-m)^\dagger. \tag{5.5}$$

The product of two tensor operators transforms under coordinate system rotations according to

$$UT_1(j_1m_1')T_2(j_2m_2')U^{-1} = (UT_1(j_1m_1')U^{-1})(UT_2(j_2m_2')U^{-1})$$
$$= \sum_{m_1m_2} T_1(j_1m_1)T_2(j_2m_2) U^{(j_1)}_{m_1m'_1} U^{(j_2)}_{m_2m'_2}. \tag{5.6}$$

It follows from (3.80) that

$$\sum_{m_1m_2} T_1(j_1m_1)T_2(j_2m_2)(j_1m_1j_2m_2|j_1j_2jm) \equiv T(j_1j_2jm) \tag{5.7}$$

obeys

$$UT(j_1j_2jm')U^{-1} = \sum_m T(j_1j_2jm)U^{(j)}_{mm'}, \tag{5.8}$$

and is therefore an irreducible tensor of rank j.

For a tensor operator applied to an angular momentum eigenvector we have, analogously,

$$U(T(j_1m_1')\Psi(j_2m_2')) = (UT(j_1m_1')U^{-1})(U\Psi(j_2m_2'))$$
$$= \sum_{m_1m_2} T(j_1m_1)\Psi(j_2m_2) U^{(j_1)}_{m_1m'_1} U^{(j_2)}_{m_2m'_2} \tag{5.9}$$

so that

$$\sum_{m_1m_2} T(j_1m_1)\Psi(j_2m_2)(j_1m_1j_2m_2|j_1j_2jm) \equiv \Phi(j_1j_2jm) \tag{5.10}$$

obeys

$$U\Phi(j_1j_2jm') = \sum_m \Phi(j_1j_2jm)U^{(j)}_{mm'} \tag{5.11}$$

---

[14] It is similarly impossible to identify the $\Psi''(jm)$ of (4.54) with $\Psi(jm)$, for all m, if j is half-integral.

and is therefore an angular momentum eigenvector with quantum numbers j and m.

The magnetic quantum number dependence of tensor operator matrix elements is contained in the last statement. On introducing explicitly the additional quantum numbers necessary to form a complete set, we are led to write

$$\sum_{qm'} T(kq)\Psi(\gamma'j'm')(kqj'm'|kj'jm) = \sum_{\gamma} \Psi(\gamma jm)(2j+1)^{-1/2}[\gamma j|T^{(k)}|\gamma'j'], \quad (5.12)$$

where we have employed different letters for the tensor operator indices in order to simplify the notation. It follows from (5.12) that[15]

$$(\gamma jm|T(kq)|\gamma'j'm') = (2j+1)^{-1/2}[\gamma j|T^{(k)}|\gamma'j'](kj'jm|kqj'm')$$
$$= (-1)^{k-j'+m}[\gamma j|T^{(k)}|\gamma'j']X(jkj'; -mqm'). \quad (5.13)$$

As an alternative derivation of the latter result,[16] we remark that

$$(\gamma jm|T(kq)|\gamma'j'm') = (U\Psi(\gamma jm), UT(kq)U^{-1}U\Psi(\gamma'j'm'))$$
$$= \sum_{m''q'm'''} (\gamma jm''|T(kq')|\gamma'j'm''')(-1)^{m-m''} U^{(j)}_{-m''-m} U^{(k)}_{q'q} U^{(j')}_{m'''m'}. \quad (5.14)$$

An integration with respect to ω then yields, according to (3.82),

$$(\gamma jm|T(kq)|\gamma'j'm') = \sum_{m''q'm'''} (-1)^{m-m''} X(jkj'; -mqm')X(jkj'; -m''q'm''')$$
$$\cdot (\gamma jm''|T(kq')|\gamma'j'm'''), \quad (5.15)$$

which is (5.13), with

$$[\gamma j|T^{(k)}|\gamma'j'] = \sum_{mqm'} (-1)^{j'-k-m} X(jkj'; -mqm')(\gamma jm|T(kq)|\gamma'j'm'). \quad (5.16)$$

According to the definition of the Hermitian conjugate tensor, we have

$$(\gamma jm|T^\dagger(kq)|\gamma'j'm') = i^{2q}(\gamma'j'm'|T(k-q)|\gamma jm)^*$$
$$= i^{2q}(-1)^{k-j+m'}[\gamma'j'|T^{(k)}|\gamma j]^* X(jkj'; -mqm'), \quad (5.17)$$

or

$$[\gamma j|T^{(k)\dagger}|\gamma'j'] = i^{2j'-2j}[\gamma'j'|T^{(k)}|\gamma j]^*, \quad (5.18)$$

in which use has been made of the X coefficient properties contained in (3.45) and (3.46). For a Hermitian tensor, this result reads

$$[\gamma j|T^{(k)}|\gamma'j'] = (-1)^{j-j'}[\gamma'j'|T^{(k)}|\gamma j]^*. \quad (5.19)$$

[15] The relation between the rectangular bracket symbol and the analogous quantity defined in R is
$$[\gamma j|T^{(k)}|\gamma'j'] = (-1)^{k+j-j'} (\gamma j\|T^{(k)}\|\gamma'j').$$

[16] This is the method employed by E. Wigner, "Gruppentheorie und ihre Anwendung auf die Quantenmechanik der Atomspektrem" (Braunschweig, 1931), p. 263.

If the tensor operators T_1 and T_2 of (5.7) refer to the same dynamical variables, we may write

$$(\gamma j m | T(k_1 k_2 k q) | \gamma' j' m') = (-1)^{k-j'+m} [\gamma j | T^{(k)}(k_1 k_2) | \gamma' j'] X(jkj'; -mqm'),$$
(5.20)

where in view of (5.16),

$$[\gamma j | T^{(k)}(k_1 k_2) | \gamma' j'] = \sum_{mqm'} (-1)^{j'-k-m} X(jkj'; -mqm')(k_1 q_1 k_2 q_2 | k_1 k_2 k q)$$

$$\cdot \sum_{\gamma'' j'' m''} (\gamma j m | T_1(k_1 q_1) | \gamma'' j'' m'')(\gamma'' j'' m'' | T_2(k_2 q_2) | \gamma' j' m'). \quad (5.21)$$

The resulting magnetic quantum number summation, involving four X coefficients, can be identified with a W coefficient,

$$[\gamma j | T^{(k)}(k_1 k_2) | \gamma' j'] = (2k+1)^{\frac{1}{2}} \sum_{\gamma'' j''} W(k_1 k_2 j j'; k j'')$$

$$\cdot [\gamma j | T_1^{(k_1)} | \gamma'' j''][\gamma'' j'' | T_2^{(k_2)} | \gamma' j']. \quad (5.22)$$

When T_1 and T_2 are tensor operators associated with different dynamical variables, so that

$$[T_1, J_2] = [T_2, J_1] = 0, \quad (5.23)$$

we have

$$(\gamma j_1 j_2 j m | T(k_1 k_2 k q) | \gamma' j_1' j_2' j' m') = (-1)^{k-j'+m} [\gamma j_1 j_2 j | T^{(k)}(k_1 k_2) | \gamma' j_1' j_2' j']$$

$$\cdot X(jkj'; -mqm'). \quad (5.24)$$

Here

$$[\gamma j_1 j_2 j | T^{(k)}(k_1 k_2) | \gamma' j_1' j_2' j'] = \sum (-1)^{j'-k-m} X(jkj'; -mqm')$$

$$\cdot (k_1 q_1 k_2 q_2 | k_1 k_2 k q)(j_1 j_2 j m | j_1 m_1 j_2 m_2)(j_1' j_2' j' m' | j_1' m_1' j_2' m_2')$$

$$\cdot (\gamma j_1 m_1 | T_1(k_1 q_1) | \gamma'' j_1'' m_1')(\gamma'' j_2 m_2 | T_2(k_2 q_2) | \gamma' j_2' m_2'), \quad (5.25)$$

where the sum is over all m variables. This magnetic quantum number summation, involving six X coefficients, can be identified with an S coefficient,

$$[\gamma j_1 j_2 j | T^{(k)}(k_1 k_2) | \gamma' j_1' j_2' j']$$
$$= [(2j+1)(2j'+1)(2k+1)]^{1/2} (-1)^{j_2+j_1'-j'-k_1+k} S(j_1 j_2 j_1' j_2'; jj' k_1 k_2; k)$$

$$\cdot \sum_{\gamma''} [\gamma j_1 | T_1^{(k_1)} | \gamma'' j_1''][\gamma'' j_2 | T_2^{(k_2)} | \gamma' j_2']. \quad (5.26)$$

Special examples which require only the W coefficient are

$$[\gamma j_1 j_2 j | T^{(0)}(k_1 k_1) | \gamma' j_1' j_2' j] = \left[\frac{2j+1}{2k_1+1}\right]^{1/2} (-1)^{j_2+j_1'-j-k_1} W(j_1 j_2 j_1' j_2'; jk_1)$$

$$\cdot \sum_{\gamma''} [\gamma j_1 | T_1^{(k_1)} | \gamma'' j_1''][\gamma'' j_2 | T_2^{(k_1)} | \gamma' j_2'], \quad (5.27)$$

$[\gamma j_1 j_2 j | T_1^{(k)} | \gamma' j_1' j_2 j']$
$= [(2j+1)(2j'+1)]^{1/2} (-1)^{j_2+k-j_1-j'} W(j_1 j j_1' j'; j_2 k) [\gamma j_1 | T_1^{(k)} | \gamma' j_1'],$ (5.28)

and

$[\gamma j_1 j_2 j | T_2^{(k)} | \gamma' j_1 j_2' j']$
$= [(2j+1)(2j'+1)]^{1/2} (-1)^{j_1+k-j'_2-j} W(j_2 j j_2' j'; j_1 k) [\gamma j_2 | T_2^{(k)} | \gamma' j_2'].$ (5.29)

Further relations connecting the S and W coefficients can be deduced from these results. We shall illustrate this for the simpler situation in which only W is involved. We multiply the two scalar operators[17]

$$T^{(0)}(k_1 k_1) = \sum_{q_1} (2k_1+1)^{-1/2} T_1(k_1 q_1)(-1)^{k_1-q_1} T_2(k_1 - q_1),$$ (5.30)

and

$$T^{(0)}(k_2 k_2) = \sum_{q_2} (2k_2+1)^{-1/2} T_1(k_2 q_2)(-1)^{k_2-q_2} T_2(k_2 - q_2),$$ (5.31)

to obtain

$$T^{(0)}(k_1 k_1) T^{(0)}(k_2 k_2) = \sum_{q_1 q_2} [(2k_1+1)(2k_2+1)]^{-1/2} T_1(k_1 q_1) T_1(k_2 q_2)$$
$$\cdot (-1)^{k_1+k_2-q_1-q_2} T_2(k_1-q_1) T_2(k_2-q_2).$$ (5.32)

On writing

$$T_1(k_1 q_1) T_1(k_2 q_2) = \sum_{kq} T_1(k_1 k_2 kq)(k_1 k_2 kq | k_1 q_1 k_2 q_2)$$ (5.33)

and

$$T_2(k_1-q_1) T_2(k_2-q_2) = \sum_{kq} T_2(k_1 k_2 k-q)(-1)^{k_1+k_2-k}(k_1 k_2 kq | k_1 q_1 k_2 q_2),$$ (5.34)

this becomes

$$T^{(0)}(k_1 k_1) T^{(0)}(k_2 k_2)$$
$$= \sum_{kq} [(2k_1+1)(2k_2+1)]^{-1/2} T_1(k_1 k_2 kq)(-1)^{k-q} T_2(k_2 k_2 k-q)$$
$$\equiv \sum_k \left[\frac{2k+1}{(2k_1+1)(2k_2+1)} \right]^{1/2} T^{(0)}([k_1 k_2] k [k_1 k_2] k).$$ (5.35)

A matrix element of this equation, when evaluated with the aid of (5.22) and (5.27), yields the information that

$$W(j_1 j_2 j_1'' j_2''; j k_1) W(j_1'' j_2'' j_1' j_2'; j k_2)$$
$$= (-1)^{j''_1+j''_2-j} \sum_k (2k+1)(-1)^{k_1+k_2-k} W(j_1 j_2 j_1' j_2'; jk)$$
$$\cdot W(k_1 k_2 j_1 j_1'; k j_1'') W(k_1 k_2 j_2 j_2'; k j_2'').$$ (5.36)

---

[17] Here T_1 and T_2 are functions of different dynamical variables.

Tensor operators can be constructed from the spin creation and annihilation operators. Thus, consider the operator

$$e^{\zeta_+(za^+)+\zeta_-[za]} = \sum_{kq\alpha} \phi_{kq}(z)\phi_{k\alpha}(\zeta)t(kq\alpha), \tag{5.37}$$

formed from the commuting quantities (za^+) and $[za]$. On subjecting this to a unitary transformation, we find

$$e^{\zeta_+(za'^+)+\zeta_-[za']} = \sum_{kq\alpha} \phi_{kq}(z)\phi_{k\alpha}(\zeta)Ut(kq\alpha)U^{-1}, \tag{5.38}$$

where the transformed creation and annihilation operators are described by (2.14). Now, according to (2.19), we have

$$(za'^+) = (z'a^+), \quad [za'] = [z'a], \quad z' = uz, \tag{5.39}$$

in which the second statement stems from the fact that a_- and a_+ transform in the same way as a_+^+ and $-a_-^+$. Therefore,

$$e^{\zeta_+(za'^+)+\zeta_-[za']} = \sum_{kq\alpha} \phi_{kq}(uz)\phi_{k\alpha}(\zeta)t(kq\alpha)$$

$$= \sum_{kqq'\alpha} U^{(k)}_{qq'}\phi_{kq'}(z)\phi_{k\alpha}(\zeta)t(kq\alpha), \tag{5.40}$$

on employing (2.21). We have thereby shown that

$$Ut(kq'\alpha)U^{-1} = \sum_q t(kq\alpha)U^{(k)}_{qq'}. \tag{5.41}$$

On taking the Hermitian conjugate of (5.37) and making the substitution $z_+^* \to z_-$, $z_-^* \to -z_+$, $\zeta_-^* \to \zeta_+$, $\zeta_+^* \to -\zeta_-$, which restores this generating operator to its original form, we find that

$$t(kq\alpha) = (-1)^{q+a}t(k-q-\alpha)^\dagger. \tag{5.42}$$

Accordingly, the adjoint tensor is given by

$$t^\dagger(kq\alpha) = i^{2\alpha}t(kq-\alpha). \tag{5.43}$$

The significance of α can be appreciated from

$$\frac{(za^+)^{k+\alpha}[za]^{k-\alpha}}{[(k+\alpha)!(k-\alpha)!]^{1/2}} = \sum_q \phi_{kq}(z)t(kq\alpha), \tag{5.44}$$

namely, 2α is the excess of creation with respect to annihilation operators. Therefore, if $t(kq\alpha)$ is applied to an angular momentum eigenvector with quantum number j', it will produce an eigenvector with quantum number j, such that

$$\alpha = j - j'. \tag{5.45}$$

To evaluate the matrix elements of $t(kq\alpha)$, we examine

$$(e^{(xa^+)}\Psi_0, e^{\zeta_+[za^+]+\zeta_-[za]} \cdot e^{(ya^+)}\Psi_0)$$
$$= \sum_{jmj'm'kq\alpha} \phi_{jm}(x^*)(jm|t(kq\alpha)|j'm')\phi_{j'm'}(y)\phi_{kq}(z)\phi_{k\alpha}(\zeta) = e^{(x^*y)+\zeta_+(x^*z)+\zeta_-[zy]}. \quad (5.46)$$

The substitution $x_+^* \to x_-$, $x_-^* \to -x_+$ places this in the form

$$\sum_{jmj'm'kq\alpha} (-1)^{j-m}\phi_{j-m}(x)(jm|t(kq\alpha)|j'm')\phi_{j'm'}(y)\phi_{kq}(z)\phi_{k\alpha}(\zeta) = e^{\zeta_-[zy]+[yx]-\zeta_+[xz]}, \quad (5.47)$$

and comparison with (3.42) shows that

$$e^{\zeta_-[zy]+[yx]-\zeta_+[xz]} = \sum_{jmj'm'kq} \phi_{j-m}(x)\phi_{kq}(z)\phi_{j'm'}(y)\phi_{k,\,j-j'}(\zeta)$$
$$\cdot (-1)^{k+j-j'}\left[\frac{(j+j'+k+1)!}{(j+j'-k)!}\right]^{1/2} X(jkj';\,-mqm'). \quad (5.48)$$

Therefore

$$(jm|t(kq\alpha)|j'm) = \delta_{\alpha,\,j-j'}(-1)^{k-j'+m}\left[\frac{(j+j'+k+1)!}{(j+j'-k)!}\right]^{1/2} X(jkj';\,-mqm), \quad (5.49)$$

or

$$[j|t^{(k)}(\alpha)|j'] = \delta_{\alpha,\,j-j'}\left[\frac{(j+j'+k+1)!}{(j+j'-k)!}\right]^{1/2}. \quad (5.50)$$

Of particular interest are the operators with $\alpha = 0$ (k integral),

$$\frac{(za)^k[za]^k}{k!} = \sum_q \phi_{kq}(z)t(kq0). \quad (5.51)$$

Indeed

$$-(za^+)[za] = \mathbf{a}\cdot\mathbf{J}, \quad (5.52)$$

where $\mathbf{a}$ is a null vector,

$$\mathbf{a}\cdot\mathbf{a} = 0, \quad (5.53)$$

with the components

$$\mathfrak{a}_1 = -z_+^2 + z_-^2, \qquad \mathfrak{a}_2 = -i(z_+^2 + z_-^2), \qquad \mathfrak{a}_3 = 2z_+z_-. \quad (5.54)$$

It is well known that if $\mathbf{r}$ is a position vector, $(\mathbf{a}\cdot\mathbf{r})^k$ is a spherical harmonic of order k,

$$\frac{(\mathbf{a}\cdot\mathbf{r})^k}{2^k k!} = \left[\frac{4\pi}{2k+1}\right]^{1/2} \sum_q \phi_{kq}(z) Y_{kq}(\mathbf{r}), \quad (5.55)$$

where $Y_{kq}(\mathbf{r})$, which usually designates a surface spherical harmonic, here includes the factor r^k. Accordingly, we write

$$\frac{(\mathbf{a}\cdot\mathbf{J})^k}{2^k k!} = \left[\frac{4\pi}{2k+1}\right]^{1/2} \sum_q \phi_{kq}(z) Y_{kq}(\mathbf{J}), \quad (5.56)$$

in which $Y_{kq}(\mathbf{J})$ differs from the analogous $Y_{kq}(\mathbf{r})$ only in that the order of factors is significant. With this notation, we have

$$t(kq0) = \left[\frac{4\pi}{2k+1}\right]^{1/2} (-2)^k Y_{kq}(\mathbf{J}), \qquad (5.57)$$

and

$$[j|\,Y^{(k)}(\mathbf{J})|j] = \left[\frac{2k+1}{4\pi}\right]^{1/2} \cdot (-\tfrac{1}{2})^k \cdot \left[\frac{(2j+k+1)!}{(2j-k)!}\right]^{1/2}. \qquad (5.58)$$

Notice also that the tensor $t(kq0)$ is Hermitian, according to (5.43), so that the operator harmonics satisfy

$$Y_{kq}(\mathbf{J})^\dagger = (-1)^q Y_{k-q}(\mathbf{J}). \qquad (5.59)$$

The matrix elements of the tensor operator

$$Y(k_1 k_2 kq) = \sum_{q_1 q_2} Y_{k_1 q_1}(\mathbf{J}) Y_{k_2 q_2}(\mathbf{J})(k_1 q_1 k_2 q_2 | k_1 k_2 kq) \qquad (5.60)$$

are described by

$$[j|\,Y^{(k)}(k_1 k_2)|j] = (2k+1)^{1/2} W(k_1 k_2 jj;\, kj)[j|\,Y^{(k_1)}|j][j|\,Y^{(k_2)}|j], \qquad (5.61)$$

in view of (5.22). With respect to their effect on an eigenvector with quantum number j, one can assert that

$$Y(k_1 k_2 kq) = Y_{kq}(\mathbf{J}) \frac{[j|\,Y^{(k)}(k_1 k_2)|j]}{[j|\,Y^{(k)}|j]}, \qquad (5.62)$$

which becomes a generally valid operator equation on replacing $j(j+1)$ with $\mathbf{J}^2$. Hence

$$\sum_{q_1 q_2} Y_{k_1 q_1}(\mathbf{J}) Y_{k_2 q_2}(\mathbf{J})(k_1 q_1 k_2 q_2 | k_1 k_2 kq)$$
$$= Y_{kq}(\mathbf{J})(2k+1)^{1/2} W(k_1 k_2 jj;\, kj) \frac{[j|\,Y^{(k_1)}|j][j|\,Y^{(k_2)}|j]}{[j|\,Y^{(k)}|j]}. \qquad (5.63)$$

The example of this result for $k=0$ can be written

$$\sum_q Y_{kq}(\mathbf{J}) Y_{kq}(\mathbf{J})^\dagger = \frac{1}{2j+1}[j|\,Y^{(k)}|j]^2 = \frac{2k+1}{4\pi} \frac{1}{4^k} \frac{(2j+k+1)!}{(2j+1)(2j-k)!} \qquad (5.64)$$

in which we have employed

$$W(kkjj;\, 0j) = (-1)^k[(2j+1)(2k+1)]^{-1/2}. \qquad (5.65)$$

One can easily exhibit the right side of (5.64) as a function of $j(j+1)$, and thus obtain the operator equation

$$\sum_q Y_{kq}(\mathbf{J}) Y_{kq}(\mathbf{J})^\dagger = \frac{2k+1}{4\pi} \{\mathbf{J}^2\}^k,$$

$$\{\mathbf{J}^2\}^k \equiv \prod_{n=0}^{k-1}\left[\mathbf{J}^2 - \frac{n}{2}\left(\frac{n}{2}+1\right)\right]. \tag{5.66}$$

The structure of the operator $\{\mathbf{J}^2\}^k$ can also be inferred from the two requirements that it annihilate any eigenvector with $j < \tfrac{1}{2}k$, and that it simplify to the kth power of $\mathbf{J}^2$ as j becomes very large.

We return to (5.63), displayed in the form

$$Y_{k_1q_1}(\mathbf{J})Y_{k_2q_2}(\mathbf{J})$$
$$= \left[\frac{(2k_1+1)(2k_2+1)}{4\pi}\right]^{1/2} \sum_{kq} Y_{kq}(\mathbf{J}) f_{k_1k_2k}(\mathbf{J}^2)(k_1k_2kq|k_1q_1k_2q_2), \tag{5.67}$$

where

$$\left[\frac{\{\mathbf{J}^2\}^k}{\{\mathbf{J}^2\}^{k_1}\{\mathbf{J}^2\}^{k_2}}\right]^{1/2} f_{k_1k_2k}(\mathbf{J}^2) = (2j+1)^{1/2} W(k_1k_2jj;kj). \tag{5.68}$$

The analogous equation for $Y_{k_2q_2}(\mathbf{J})Y_{k_1q_1}(\mathbf{J})$ differs from (5.67) only by the inclusion of the factor $(-1)^{k_1+k_2-k}$, as follows from (3.47). The addition and subtraction of these two equations then yields

$$\{Y_{k_1q_1}(\mathbf{J}), Y_{k_2q_2}(\mathbf{J})\} = \left[\frac{(2k_1+1)(2k_2+1)}{4\pi}\right]^{1/2}$$
$$\cdot \sum_{\text{even},\,kq} Y_{kq}(\mathbf{J}) f_{k_1k_2k}(\mathbf{J}^2)(k_1k_2kq|k_1q_1k_2q_2) \tag{5.69}$$

and

$$[Y_{k_1q_1}(\mathbf{J}), Y_{k_2q_2}(\mathbf{J})] = \left[\frac{(2k_1+1)(2k_2+1)}{4\pi}\right]^{1/2}$$
$$\cdot \sum_{\text{odd},\,kq} Y_{kq}(\mathbf{J}) f_{k_1k_2k}(\mathbf{J}^2)(k_1k_2kq|k_1q_1k_2q_2), \tag{5.70}$$

where the parity referred to is that of $k_1 + k_2 - k$. In the latter equation we have the commutation properties of these operator functions of $\mathbf{J}$.

As an elementary application of (5.70), we take its trace for the states with quantum number j. In view of the null trace possessed by a commutator, we infer that the trace of $Y_{kq}(\mathbf{J})$ vanishes for every k that can occur in (5.70). Since these k values are $|k_1 - k_2| + 1$, $|k_1 - k_2| + 3$, ..., $k_1 + k_2 - 1$, we obtain[18]

$$\text{tr}^{(j)} Y_{kq}(\mathbf{J}) = 0, \qquad k > 0, \tag{5.71}$$

[18] This theorem is easily proved for an arbitrary tensor operator by taking the trace of (5.1) for states with a given j, and integrating with respect to ω

$$\sum_{m=-j}^{j} (\gamma jm|T(kq)|\gamma' jm) = 0, \qquad k > 0.$$

Of course, k must be integral if the individual matrix elements are not to vanish.

or
$$\frac{1}{2j+1}\operatorname{tr}^{(j)} Y_{kq}(\mathbf{J}) = \delta_{k0}. \tag{5.72}$$

With the aid of this result, the trace of (5.67) is evaluated as

$$\frac{1}{2j+1}\operatorname{tr}^{(j)} Y_{k_1q_1}(\mathbf{J})^\dagger Y_{k_2q_2}(\mathbf{J}) = \frac{1}{4\pi}\{j(j+1)\}^{k_1}\delta_{k_1k_2}\delta_{q_1q_2}, \tag{5.73}$$

which expresses the orthogonality of the operator harmonics. The multiplication of (5.67) with $Y_{k_3q_3}(\mathbf{J})$ then yields

$$\frac{1}{2j+1}\operatorname{tr}^{(j)} Y_{k_1q_1}(\mathbf{J}) Y_{k_2q_2}(\mathbf{J}) Y_{k_3q_3}(\mathbf{J}) = \left[\prod_i \frac{2k_i+1}{4\pi}\{j(j+1)\}^{k_i}\right]^{1/2}$$
$$\cdot X(k;q)(-1)^{k_1-k_2}(2j+1)^{1/2}W(k_1k_2jj;k_3j). \tag{5.74}$$

A comparison with (3.85) shows that, in the limit of large j,

$$(-1)^{k_1-k_2}(2j+1)^{1/2}W(k_1k_2jj;k_3j) \to X(k_1k_2k_3;000). \tag{5.75}$$

Turning to tensor operators formed from two angular momenta, we remark that, for matrix elements diagonal in j,

$$\sum_{q_1q_2} Y_{k_1q_1}(\mathbf{J}_1) Y_{k_2q_2}(\mathbf{J}_2)(k_1q_1k_2q_2|k_1k_2kq)$$
$$= Y_{kq}(\mathbf{J})(2j+1)(2k+1)^{1/2}(-1)^{j_1+j_2-j-k_1+k}S(j_1j_2j_1j_2;jjk_1k_2;k)$$
$$\cdot \frac{[j_1|Y^{(k_1)}|j_1][j_2|Y^{(k_2)}|j_2]}{[j|Y^{(k)}|j]}. \tag{5.76}$$

No such restriction is required for the special example

$$\sum_q Y_{kq}(\mathbf{J}_1)^\dagger Y_{kq}(\mathbf{J}_2) = (-1)^{j_1+j_2-j}W(j_1j_2j_1j_2;jk)[j_1|Y^{(k)}|j_1][j_2|Y^{(k)}|j_2]. \tag{5.77}$$

In terms of the Legendre polynomial operator defined by

$$\sum_q Y_{kq}(\mathbf{J}_1)^\dagger Y_{kq}(\mathbf{J}_2) = \frac{2k+1}{4\pi} P_k(\mathbf{J}_1,\mathbf{J}_2), \tag{5.78}$$

the latter equation can be written

$$[\{\mathbf{J}_1^2\}^k\{\mathbf{J}_2^2\}^k]^{-1/2} P_k(\mathbf{J}_1,\mathbf{J}_2) = (-1)^{j_1+j_2-j}(2j_1+1)^{1/2}(2j_2+1)^{1/2}W(j_1j_2j_1j_2;jk), \tag{5.79}$$

which indicates the limiting form of the right side for large j_1, j_2, and j. The simple result obtained for $j=0$ can be expressed as

$$P_k(\mathbf{J},-\mathbf{J}) = (-1)^k\{\mathbf{J}^2\}^k. \tag{5.80}$$

A multiplication theorem for the Legendre operator is obtained from the observation that

$$\frac{2k_1+1}{4\pi}\frac{2k_2+1}{4\pi}P_{k_1}(\mathbf{J}_1,\mathbf{J}_2)P_{k_2}(\mathbf{J}_1,\mathbf{J}_2)$$
$$=\sum_{k,q}(Y_{k_2q_2}(\mathbf{J}_1)Y_{k_1q_1}(\mathbf{J}_1))^\dagger Y_{k_1q_1}(\mathbf{J}_2)Y_{k_2q_2}(\mathbf{J}_2), \quad (5.81)$$

namely,[19]

$$P_{k_1}(\mathbf{J}_1,\mathbf{J}_2)P_{k_2}(\mathbf{J}_1,\mathbf{J}_2)$$
$$=\sum_k(2k+1)P_k(\mathbf{J}_1,\mathbf{J}_2)(-1)^{k_1+k_2-k}f_{k_1k_2k}(\mathbf{J}_1^2)f_{k_1k_2k}(\mathbf{J}_2^2). \quad (5.82)$$

On placing $k_2=1$, we obtain a simple recurrence relation from which the Legendre operators can be constructed successively, starting with

$$P_0(\mathbf{J}_1,\mathbf{J}_2)=1. \quad (5.83)$$

The coefficients in the recurrence relation can be computed from

$$W(k1jj;k+1\,j)=-W(k+1\,1jj;kj)$$
$$=\left[\frac{4j(j+1)-k^2-2k}{4j(j+1)(2j+1)}\frac{k+1}{(2k+1)(2k+3)}\right]^{1/2}, \quad (5.84)$$

and

$$\sum_k(2k+1)(2j+1)[W(k_1 1jj;kj)]^2=1. \quad (5.85)$$

Thus

$$f_{k1k+1}(\mathbf{J}^2)=\left[\frac{k+1}{(2k+1)(2k+3)}\right]^{1/2},$$
$$f_{k1k-1}(\mathbf{J}^2)=-\left[\frac{k}{(2k-1)(2k+1)}\right]^{1/2}\left(\mathbf{J}^2-\frac{k^2-1}{4}\right),$$
$$(f_{k1k}(\mathbf{J}^2))^2=\tfrac{1}{4}\frac{k(k+1)}{2k+1}, \quad (5.86)$$

and therefore,

$$\left(\mathbf{J}_1\cdot\mathbf{J}_2+\frac{k(k+1)}{4}\right)P_k(\mathbf{J}_1,\mathbf{J}_2)=\frac{k+1}{2k+1}P_{k+1}(\mathbf{J}_1,\mathbf{J}_2)$$
$$+\frac{k}{2k+1}\left(\mathbf{J}_1^2-\frac{k^2-1}{4}\right)\left(\mathbf{J}_2^2-\frac{k^2-1}{4}\right)P_{k-1}(\mathbf{J}_1,\mathbf{J}_2). \quad (5.87)$$

---

[19] This is a particular example of the theorem on the product of two W coefficients [Eq. (5.36)].

As the first few Legendre operators, obtained in succession from (5.87) with $k = 0, 1, 2$, we have

$$P_1(\mathbf{J}_1, \mathbf{J}_2) = \mathbf{J}_1 \cdot \mathbf{J}_2,$$

$$P_2(\mathbf{J}_1, \mathbf{J}_2) = \tfrac{3}{2}\mathbf{J}_1 \cdot \mathbf{J}_2(\mathbf{J}_1 \cdot \mathbf{J}_2 + \tfrac{1}{2}) - \tfrac{1}{2}\mathbf{J}_1^2 \mathbf{J}_2^2,$$

$$P_3(J_1, J_2) = \tfrac{5}{2}\mathbf{J}_1 \cdot \mathbf{J}_2(\mathbf{J}_1 \cdot \mathbf{J}_2 + \tfrac{1}{2})(\mathbf{J}_1 \cdot \mathbf{J}_2 + \tfrac{3}{2}) - \tfrac{5}{6}(\mathbf{J}_1 \cdot \mathbf{J}_2 + \tfrac{3}{2})\mathbf{J}_1^2\mathbf{J}_2^2$$
$$- \tfrac{2}{3}\mathbf{J}_1 \cdot \mathbf{J}_2(\mathbf{J}_1^2 - \tfrac{3}{4})(\mathbf{J}_2^2 - \tfrac{3}{4}). \tag{5.88}$$

A useful check upon these results is afforded by (5.80).

A statement analogous to (5.62) can be made for an arbitrary tensor operator; as far as matrix elements diagonal in j are concerned,

$$T(kq) = Y_{kq}(\mathbf{J}) \frac{[j|T^{(k)}|j]}{[j|Y^{(k)}|j]}. \tag{5.89}$$

The coefficient in this relation can be expressed in other ways. Thus, we have

$$\sum_q Y_{kq}(\mathbf{J})^\dagger T(kq) = \sum_q Y_{kq}(\mathbf{J})^\dagger Y_{kq}(\mathbf{J}) \frac{[j|T^{(k)}|j]}{[j|Y^{(k)}|j]}, \tag{5.90}$$

which leads to the projection rule

$$T(kq) \to \frac{4\pi}{2k+1} Y_{kq}(\mathbf{J}) \frac{1}{\{\mathbf{J}^2\}^k} \sum_{q'} Y_{kq'}(\mathbf{J})^\dagger T(kq'), \tag{5.91}$$

for isolating the part of a tensor operator that contributes to matrix elements diagonal in j. Alternatively, we consider the particular matrix element

$$(jj|T(k0)|jj) = (jj|Y_{k0}(\mathbf{J})|jj) \frac{[j|T^{(k)}|j]}{[j|Y^{(k)}|j]}. \tag{5.92}$$

Now

$$(jj|Y_{k0}(\mathbf{J})|jj) = (-1)^k [j|Y^{(k)}|j] X(jkj; -j0j) = \left[\frac{2k+1}{4\pi}\right]^{1/2} \frac{1}{2^k} \frac{(2j)!}{(2j-k)!}, \tag{5.93}$$

so that, for matrix elements diagonal in j,

$$T(kq) = \left[\frac{4\pi}{(2k+1)}\right]^{1/2} Y_{kq}(\mathbf{J}) 2^k \frac{(2j-k)!}{(2j)!} (jj|T(k0)|jj). \tag{5.94}$$

Appendix A

We shall describe a method which produces simultaneously the eigenvalues and eigenvectors of the angular momentum operators. Consider for this purpose the unitary operator

$$V = \exp(i\chi \tfrac{1}{2}n + i\phi J_3), \tag{A1}$$

which has the eigenvalues $\exp(ij\chi + im\phi)$. The operator V can be interpreted as

$$V = \sum_{jm} [\exp(ij\chi + im\phi)] P(jm), \tag{A2}$$

where $P(jm)$, the projection operator for the state with the indicated eigenvalues, is represented in terms of the corresponding eigenvector by the dyadic

$$P(jm) = \Psi(jm)\Psi(jm)^*. \tag{A3}$$

Accordingly, if V can be constructed and displayed in the form (A2), we shall have achieved our goal.

We write

$$V = \exp(\tfrac{1}{2}i(\gamma_+ n_+ + \gamma_- n_-)),$$
$$\gamma_+ = \chi + \phi, \qquad \gamma_- = \chi - \phi, \tag{A4}$$

and deduce the differential equations

$$\frac{\partial}{\partial \gamma_\zeta} V = \tfrac{1}{2} i a_\zeta^+ a_\zeta V$$
$$= \tfrac{1}{2} i [\exp(\tfrac{1}{2} i \gamma_\zeta)] a_\zeta^+ V a_\zeta, \tag{A5}$$

with the aid of

$$V^{-1} a_\zeta V = (\exp(\tfrac{1}{2} i \gamma_\zeta)) a_\zeta. \tag{A6}$$

The latter can be verified by differentiation,

$$\frac{\partial}{\partial \gamma_\zeta} V^{-1} a_\zeta V = \tfrac{1}{2} i V^{-1} [a_\zeta, n_\zeta] V = \tfrac{1}{2} i V^{-1} a_\zeta V, \tag{A7}$$

or from the general theorem

$$a_\zeta f(n_\zeta) = f(n_\zeta + 1) a_\zeta. \tag{A8}$$

In virtue of the operator ordering in (A5), the solution of these equations which reduces to unity for $\gamma_\zeta = 0$ is given by

$$V = \exp\{(e^{\tfrac{1}{2} i \gamma_+} - 1) a_+^+; a_+ + (e^{\tfrac{1}{2} i \gamma_-} - 1) a_-^+; a_-\} \tag{A9}$$

where

$$\exp(\lambda a^+; a) = \sum \frac{\lambda^n}{n!} (a^+)^n (a)^n \tag{A10}$$

is a correspondingly ordered form of the exponential. We write this solution as

$$V = \exp(\sum_\zeta e^{\tfrac{1}{2} i \gamma_\zeta} a_\zeta^+; P_0; a_\zeta), \tag{A11}$$

Appendix A

which is intended to indicate that

$$P_0 = \exp(-a_+^+; a_+ - a_-^+; a_-) = \exp(-(a^+; a)) \qquad (A12)$$

is to be inserted between the powers of a_ζ^+ and a_ζ in the ordered operator expansion of V:

$$V = \sum_{n_+, n_- = 0}^{\infty} \{\exp[\tfrac{1}{2}i(n_+ \gamma_+ + n_- \gamma_-)]\} \frac{(a_+^+)^{n_+}(a_-^+)^{n_-}}{(n_+! n_-!)^{1/2}} P_0 \frac{(a_+)^{n_+}(a_-)^{n_-}}{(n_+! n_-!)^{1/2}}. \qquad (A13)$$

We have thus obtained the form (A2), with

$$j = \tfrac{1}{2}(n_+ + n_-), \qquad m = \tfrac{1}{2}(n_+ - n_-), \qquad n_+, n_- = 0, 1, 2, \ldots, \qquad (A14)$$

and

$$P(jm) = \phi_{jm}(a^+) P_0 \phi_{jm}(a), \qquad (A15)$$

in which we have employed the notation

$$\phi_{jm}(a^+) = \frac{(a_+^+)^{j+m}(a_-^+)^{j-m}}{[(j+m)!(j-m)!]^{1/2}}. \qquad (A16)$$

In terms of the eigenvector Ψ_0, defined by

$$P_0 = \Psi_0 \Psi_0^*, \qquad (A17)$$

the angular momentum eigenvectors are exhibited as

$$\Psi(jm) = \phi_{jm}(a^+) \Psi_0. \qquad (A18)$$

The fundamental property of $\Psi_0 = \Psi(00)$ is deduced from

$$[a_\zeta, P_0] = (\partial/\partial a_\zeta^+) P_0 = -P_0 a_\zeta, \qquad (A19)$$

or

$$a_\zeta P_0 = 0, \qquad (A20)$$

namely

$$a_\zeta \Psi_0 = 0. \qquad (A21)$$

The simple generating function for the eigenvectors, (1.16), can also be obtained by noting that

$$(\Psi(jm), e^{(xa^+)} \Psi_0) = (\Psi_0, \phi_{jm}(a) e^{(xa^+)} \Psi_0)$$
$$= \phi_{jm}(x). \qquad (A22)$$

Indeed,

$$e^{(xa^+)} \Psi_0 = \sum_{jm} \Psi(jm)(\Psi(jm), e^{(xa^+)} \Psi_0)$$
$$= \sum_{jm} \phi_{jm}(x) \Psi(jm). \qquad (A23)$$

Appendix B

The ordered operator

$$A = \exp(za; a^+), \quad [a, a^+] = 1, \tag{B1}$$

satisfies

$$[a, A] = (\partial/\partial a^+)A = zaA,$$

$$[a^+, A] = -(\partial/\partial a)A = -Aza^+, \tag{B2}$$

or

$$(1-z)aA = Aa, \quad a^+A = (1-z)Aa^+. \tag{B3}$$

Therefore

$$\frac{\partial}{\partial z}A = aAa^+ = \frac{1}{1-z}Aaa^+ = \frac{1}{1-z}A + \frac{1}{1-z}Aa^+a$$

$$= \frac{1}{1-z}A + \frac{1}{(1-z)^2}a^+Aa, \tag{B4}$$

the solution of which implies the ordered operator identity,

$$\exp(za; a^+) = \frac{1}{1-z}\exp\left(\frac{z}{1-z}a^+; a\right). \tag{B5}$$

A particular consequence of this relation

$$\exp(za; a^+) \cdot \Psi_0 = \frac{1}{1-z}\Psi_0, \quad a\Psi_0 = 0, \tag{B6}$$

is derived directly in the text (Eq. (2.35)). The properties of A contained in (B3) are also displayed in the generalizations

$$\exp(za; a^+) \cdot f(a^+) = \frac{1}{1-z} f\left(\frac{a^+}{1-z}\right) \exp\left(\frac{z}{1-z}a^+; a\right),$$

$$f(a) \exp(za; a^+) = \exp\left(\frac{z}{1-z}a^+; a\right) \frac{1}{1-z} f\left(\frac{a}{1-z}\right). \tag{B7}$$

The particular examples of these identities provided by

$$\exp(za; a^+) \cdot (a^+)^r = \frac{(a^+)^r}{(1-z)^{r+1}} \exp\left(\frac{z}{1-z}a^+; a\right) \tag{B8}$$

and

$$a^r \exp(za; a^+) = \exp\left(\frac{z}{1-z}a^+; a\right) \frac{a^r}{(1-z)^{r+1}} \tag{B9}$$

are operator forms of the Laguerre polynomial generating functions. Thus, if we place $a^+ = x$, $a = \partial/\partial x$, and let both sides of (B8) operate upon e^{-x}, we obtain

$$\sum_{n=0}^{\infty} \frac{z^n}{n!} \left(\frac{\partial}{\partial x}\right)^n x^{n+r} e^{-x} = \frac{x^r}{(1-z)^{r+1}} \exp\left(\frac{z}{1-z} x; \frac{\partial}{\partial x}\right) e^{-x}$$

$$= \frac{x^r}{(1-z)^{r+1}} \exp\left(-\frac{z}{1-z} x\right) e^{-x}, \quad (B10)$$

or

$$\frac{\exp\left(-\frac{zx}{1-z}\right)}{(1-z)^{r+1}} = \sum_{n=0}^{\infty} z^n L_n^{(r)}(x), \quad (B11)$$

where

$$L_n^{(r)}(x) = \frac{1}{n!} x^{-r} e^x \left(\frac{d}{dx}\right)^n (x^{n+r} e^{-x}). \quad (B12)$$

A similar procedure applied to (B9) yields

$$\sum_{n=0}^{\infty} \frac{z^n}{n!} \left(\frac{\partial}{\partial x}\right)^{n+r} x^n e^{-x} = \exp\left(\frac{z}{1-z} x; \frac{\partial}{\partial x}\right) \frac{\left(\frac{\partial}{\partial x}\right)^r}{(1-z)^{r+1}} e^{-x}$$

$$= \frac{(-1)^r}{(1-z)^{r+1}} \exp\left(-\frac{z}{1-z} x\right) e^{-x}, \quad (B13)$$

which proves the equivalence between (B12) and

$$L_n^{(r)}(x) = \frac{(-1)^r}{n!} e^x \left(\frac{d}{dx}\right)^{n+r} (x^n e^{-x}). \quad (B14)$$

Another example of an ordered operator identity involves the cylinder function [Eq. (1.36)]

$$F_r(z) = r! z^{-r/2} I_r(2z^{1/2}) = \frac{r!}{2\pi i} \oint dt \frac{\exp\left(t + \frac{z}{t}\right)}{t^{r+1}}. \quad (B15)$$

We have

$$a^r F_r(za; a^+) = \frac{r!}{2\pi i} \oint dt \frac{e^t}{t^{r+1}} a^r \exp\left(\frac{z}{t} a; a^+\right)$$

$$= \frac{r!}{2\pi i} e^z \oint dt \frac{e^{t-z}}{(t-z)^{r+1}} \exp\left(\frac{z}{t-z} a^+; a\right) a^r$$

$$= e^z F_r(za^+; a) a^r, \quad (B16)$$

and similarly

$$F_r(za; a^+)(a^+)^r = e^z (a^+)^r F_r(za^+; a). \quad (B17)$$

From these identities we obtain the Laguerre polynomial generating function

$$\sum_{n=0}^{\infty} \frac{r!}{(n+r)!} z^n L_n^{(r)}(x) = e^z F_r(-zx). \tag{B18}$$

Appendix C

It is our purpose in this section to evaluate a class of scalar products, the simplest illustration of which is

$$T^{(2)} = (\exp(\lambda[a^+b^+] + \xi_+(xa^+) + \xi_-(xb^+))\Psi_0',$$
$$\exp(\kappa[a^+b^+] + \eta_+(ya^+) + \eta_-(yb^+))\Psi_0'). \tag{C1}$$

Differentiation with respect to ξ_+^* yields

$$(\partial/\partial\xi_+^*)T^{(2)} = (e\cdots\Psi_0', (x^*a)e\cdots\Psi_0')$$
$$= \eta_+(x^*y)T^{(2)} + \kappa([xb]e\cdots\Psi_0', e\cdots\Psi_0'), \tag{C2}$$

or

$$(1 - \lambda^*\kappa)(\partial/\partial\xi_+^*)T^{(2)} = \eta_+(x^*y)T^{(2)}. \tag{C3}$$

The solution of this, and analogous equations, is

$$T^{(2)} = \exp\left(\frac{(\xi^*\eta)(x^*y)}{1 - \lambda^*\kappa}\right) \cdot T_0^{(2)}, \tag{C4}$$

where

$$T_0^{(2)} = (\exp(\lambda[a^+b^+])\Psi_0', \exp(\kappa[a^+b^+])\Psi_0')$$
$$= (\Psi_0', \exp(\lambda^*\kappa(a^+;a))\Psi_0') = \frac{1}{(1 - \lambda^*\kappa)^2}, \tag{C5}$$

in view of the simple generalization of (B6)

$$\exp(z(a;a^+))\Psi_0' = \frac{1}{(1-z)^2}\Psi_0'. \tag{C6}$$

Therefore

$$T^{(2)} = \frac{1}{(1 - \lambda^*\kappa)^2} \exp\left(\frac{(\xi^*\eta)(x^*y)}{1 - \lambda^*\kappa}\right). \tag{C7}$$

One can prove, in a similar manner, that

$$(\exp(\lambda[a^+b^+] + (x_1a^+) + (x_2b^+))\Psi_0', \exp(\kappa[a^+b^+] + (y_1a^+) + (y_2b^+))\Psi_0')$$
$$= \frac{1}{(1 - \lambda^*\kappa)^2} \exp\left(\frac{1}{1 - \lambda^*\kappa}\{(x_1^*y_1) + (x_2^*y_2) + \kappa[x_1^*x_2^*] + \lambda^*[y_1y_2]\}\right). \tag{C8}$$

Appendix C

The general member of the class exemplified by (C1) is

$$T^{(n)} = (\exp(\tfrac{1}{2}\sum_{\mu\nu}\lambda_{\mu\nu}[A_\mu^+ A_\nu^+] + \sum_\mu \xi_\mu(xA_\mu^+))\Psi'_0,$$
$$\exp(\tfrac{1}{2}\sum_{\mu\nu}\kappa_{\mu\nu}[A_\mu^+ A_\nu^+] + \sum_\mu \eta_\mu(yA_\mu^+))\Psi'_0), \quad (C9)$$

where the A_μ are n sets of two-component operators, obeying

$$[A_{\zeta\mu}, A_{\zeta'\nu}^+] = \delta_{\mu\nu}\delta_{\zeta\zeta'}, \quad (C10)$$

while $\lambda_{\mu\nu}$ and $\kappa_{\mu\nu}$ form antisymmetrical matrices. Following the same procedure, we evaluate

$$(\partial/\partial\xi_\mu^*)T^{(n)} = (e\cdots\Psi'_0, (x^*A_\mu)e\cdots\Psi'_0)$$
$$= \eta_\mu(x^*y)T^{(n)} + \sum_\nu \kappa_{\mu\nu}([xA_\nu]e\cdots\Psi'_0, e\cdots\Psi'_0), \quad (C11)$$

whence

$$(\partial/\partial\xi_\mu^*)T^{(n)} + \sum_\beta \kappa_{\mu\beta}\lambda_{\beta\nu}^*(\partial/\partial\xi_\nu^*)T^{(n)} = \eta_\mu(x^*y)T^{(n)}. \quad (C12)$$

The solution of this equation can be expressed in a matrix notation as

$$T^{(n)} = \exp\left[(x^*y)\sum_{\mu\nu}\xi_\mu^*\left(\frac{1}{1+\kappa\lambda^*}\right)_{\mu\nu}\eta_\nu\right]T_0^{(n)}, \quad (C13)$$

where

$$T_0^{(n)} = (\exp(\tfrac{1}{2}\sum_{\mu\nu}\lambda_{\mu\nu}[A_\mu^+ A_\nu^+])\Psi'_0, \exp(\tfrac{1}{2}\sum_{\mu\nu}\kappa_{\mu\nu}[A_\mu^+ A_\nu^+])\Psi'_0)$$
$$= (\Psi'_0, Q\Psi'_0), \quad (C14)$$

and

$$Q = \exp(\tfrac{1}{2}\sum_{\mu\nu}\lambda_{\mu\nu}^*[A_\mu A_\nu]) \cdot \exp(\tfrac{1}{2}\sum_{\mu\nu}\kappa_{\mu\nu}[A_\mu^+ A_\nu^+]). \quad (C15)$$

To evaluate $T_0^{(n)}$, we employ the following properties of Q,

$$(\partial/\partial\lambda_{\mu\nu}^*)Q = [A_\mu A_\nu]Q, \quad (C16)$$

and

$$[xA_\mu]Q - Q[xA_\mu] = -Q\sum_\nu \kappa_{\mu\nu}(xA_\nu^+), \quad (C17a)$$

$$Q(xA_\mu^+) - (xA_\mu^+)Q = \sum_\nu \lambda_{\mu\nu}^*[xA_\nu]Q, \quad (C17b)$$

in which x is an arbitrary constant spinor. One can combine (C17a, b) into

$$\sum_\nu (1+\kappa\lambda^*)_{\mu\nu}[xA_\nu]Q = Q[xA_\mu] - \sum_\beta \kappa_{\mu\beta}(xA_\beta^+)Q, \quad (C18)$$

or

$$[xA_\nu]Q = \sum_\beta \left(\frac{1}{1+\kappa\lambda^*}\right)_{\nu\beta} Q[xA_\beta] - \sum_{\nu\beta}\left(\frac{1}{1+\kappa\lambda^*}\kappa\right)_{\nu\beta}(xA_\beta^+)Q. \quad (C19)$$

Therefore

$$[A_\mu A_\nu]Q = \sum_\beta \left(\frac{1}{1+\kappa\lambda^*}\right)_{\nu\beta} [A_\mu Q A_\beta] - \sum_\beta \left(\frac{1}{1+\kappa\lambda^*}\right)_{\nu\beta} (A_\beta^+ A_\mu)Q$$
$$- 2\left(\frac{1}{1+\kappa\lambda^*}\kappa\right)_{\nu\mu} Q, \quad (C20)$$

from which we obtain ($A_\mu \Psi_0 = 0$)

$$(\partial/\partial\lambda_{\mu\nu}^*) T_0^{(n)} = -2\left(\frac{1}{1+\kappa\lambda^*}\kappa\right)_{\nu\mu} T_0^{(n)}. \quad (C21)$$

Thus, with respect to changes in the matrix λ^*, we have

$$\delta \log T_0^{(n)} = \tfrac{1}{2}\sum_{\mu\nu} \delta\lambda_{\mu\nu}^* (\partial/\partial\lambda_{\mu\nu}^*) \log T_0^{(n)} = -\mathrm{tr}\left(\frac{1}{1+\kappa\lambda^*}\kappa\delta\lambda^*\right). \quad (C22)$$

On comparing this with the theorem on differentiation of a determinant,

$$\delta \log |A| = \mathrm{tr}(A^{-1}\delta A), \quad (C23)$$

we obtain the desired general evaluation,

$$T_0^{(n)} = \frac{1}{|1+\kappa\lambda^*|}. \quad (C24)$$

A recurrence relation for $T_0^{(n)}$ can also be established with the aid of (C13). Thus, we have

$$T_0^{(n)} = (\exp(\tfrac{1}{2}\sum_{\mu\nu=1}^{n-1} \lambda_{\mu\nu}[A_\mu^+ A_\nu^+] + \sum_{\mu\nu=1}^{n-1} \lambda_{n\mu}[A_n^+ A_\mu^+])\Psi_0,$$

$$\exp(\tfrac{1}{2}\sum_{\mu\nu=1}^{n-1} \kappa_{\mu\nu}[A_\mu^+ A_\nu^+] + \sum_{\mu\nu=1}^{n-1} \kappa_{n\mu}[A_n^+ A_\mu^+])\Psi_0)$$

$$= \left(\Psi_0, \exp\left((A_n; A_n^+)\sum_{\mu\nu} \lambda_{n\mu}^* \left(\frac{1}{1+\kappa'\lambda'^*}\right)_{\mu\nu}\kappa_{n\nu}\right)\Psi_0\right) T_0^{(n-1)}$$

$$= \left[1 - \sum_{\mu\nu} \lambda_{n\mu}^*\left(\frac{1}{1+\kappa'\lambda'^*}\right)_{\mu\nu}\kappa_{n\nu}\right]^{-2} T_0^{(n-1)}, \quad (C25)$$

in which κ' and λ' designate the matrices of dimensionality $n-1$.

The actual construction of the $T_0^{(n)}$ can be performed without detailed calculations. It follows from (C24) and (C25) that $T_0^{(n)}$ has the form of the inverse square of a power series in the components of λ^* and κ, where the last terms of the series, $(-1)^{\frac{1}{2}n}|\lambda^*|^{1/2}|\kappa|^{1/2}$, vanishes for n odd. Thus, beginning with

$$T_0^{(2)} = [1 - \lambda_{12}^*\kappa_{12}]^{-2} = [1 - |\lambda^*|^{1/2}|\kappa|^{1/2}]^{-2}, \quad (C26)$$

We infer that $T_0^{(3)}$ has the same structure, suitably extended for the additional dimension,

$$T_0^{(3)} = [1 - (\lambda_{12}^* \kappa_{12} + \lambda_{23}^* \kappa_{23} + \lambda_{31}^* \kappa_{31})]^{-2}$$

$$= [1 - \tfrac{1}{2} \sum_{\mu\nu=1}^{3} \lambda_{\mu\nu}^* \kappa_{\mu\nu}]^{-2}, \tag{C27}$$

and therefore

$$T_0^{(4)} = [1 - \tfrac{1}{2} \sum_{\mu\nu=1}^{4} \lambda_{\mu\nu}^* \kappa_{\mu\nu} + |\lambda^*|^{1/2}|\kappa|^{1/2}]^{-2}, \tag{C28}$$

where

$$|\lambda|^{1/2} = \lambda_{12}\lambda_{34} + \lambda_{23}\lambda_{14} + \lambda_{31}\lambda_{24}$$

$$= \tfrac{1}{8} \sum_{\mu\nu\sigma\tau=1}^{4} \varepsilon_{\mu\nu\sigma\tau} \lambda_{\mu\nu} \lambda_{\sigma\tau}, \tag{C29}$$

and ε is the completely antisymmetric symbol. For the last indication of this general procedure we remark that, as the extension of (C28), we have

$$T_0^{(5)} = [1 - \tfrac{1}{2} \sum_{\mu\nu=1}^{5} \lambda_{\mu\nu}^* \kappa_{\mu\nu} + \sum_{\mu\nu=1}^{5} (\lambda^*)_\alpha (\kappa)_\alpha]^{-2}, \tag{C30}$$

in which

$$(\lambda)_\alpha = \tfrac{1}{8} \sum_{\mu\nu\sigma\tau=1}^{5} \varepsilon_{\alpha\mu\nu\sigma\tau} \lambda_{\mu\nu} \lambda_{\sigma\tau}. \tag{C31}$$

A CATALOG OF SELECTED
DOVER BOOKS
IN SCIENCE AND MATHEMATICS

CATALOG OF DOVER BOOKS

Physics

THEORETICAL NUCLEAR PHYSICS, John M. Blatt and Victor F. Weisskopf. An uncommonly clear and cogent investigation and correlation of key aspects of theoretical nuclear physics by leading experts: the nucleus, nuclear forces, nuclear spectroscopy, two-, three- and four-body problems, nuclear reactions, beta-decay and nuclear shell structure. 896pp. 5 3/8 x 8 1/2. 0-486-66827-4

QUANTUM THEORY, David Bohm. This advanced undergraduate-level text presents the quantum theory in terms of qualitative and imaginative concepts, followed by specific applications worked out in mathematical detail. 655pp. 5 3/8 x 8 1/2. 0-486-65969-0

ATOMIC PHYSICS AND HUMAN KNOWLEDGE, Niels Bohr. Articles and speeches by the Nobel Prize–winning physicist, dating from 1934 to 1958, offer philosophical explorations of the relevance of atomic physics to many areas of human endeavor. 1961 edition. 112pp. 5 3/8 x 8 1/2. 0-486-47928-5

COSMOLOGY, Hermann Bondi. A co-developer of the steady-state theory explores his conception of the expanding universe. This historic book was among the first to present cosmology as a separate branch of physics. 1961 edition. 192pp. 5 3/8 x 8 1/2. 0-486-47483-6

LECTURES ON QUANTUM MECHANICS, Paul A. M. Dirac. Four concise, brilliant lectures on mathematical methods in quantum mechanics from Nobel Prize-winning quantum pioneer build on idea of visualizing quantum theory through the use of classical mechanics. 96pp. 5 3/8 x 8 1/2. 0-486-41713-1

THE PRINCIPLE OF RELATIVITY, Albert Einstein and Frances A. Davis. Eleven papers that forged the general and special theories of relativity include seven papers by Einstein, two by Lorentz, and one each by Minkowski and Weyl. 1923 edition. 240pp. 5 3/8 x 8 1/2. 0-486-60081-5

PHYSICS OF WAVES, William C. Elmore and Mark A. Heald. Ideal as a classroom text or for individual study, this unique one-volume overview of classical wave theory covers wave phenomena of acoustics, optics, electromagnetic radiations, and more. 477pp. 5 3/8 x 8 1/2. 0-486-64926-1

THERMODYNAMICS, Enrico Fermi. In this classic of modern science, the Nobel Laureate presents a clear treatment of systems, the First and Second Laws of Thermodynamics, entropy, thermodynamic potentials, and much more. Calculus required. 160pp. 5 3/8 x 8 1/2. 0-486-60361-X

QUANTUM THEORY OF MANY-PARTICLE SYSTEMS, Alexander L. Fetter and John Dirk Walecka. Self-contained treatment of nonrelativistic many-particle systems discusses both formalism and applications in terms of ground-state (zero-temperature) formalism, finite-temperature formalism, canonical transformations, and applications to physical systems. 1971 edition. 640pp. 5 3/8 x 8 1/2. 0-486-42827-3

QUANTUM MECHANICS AND PATH INTEGRALS: Emended Edition, Richard P. Feynman and Albert R. Hibbs. Emended by Daniel F. Styer. The Nobel Prize–winning physicist presents unique insights into his theory and its applications. Feynman starts with fundamentals and advances to the perturbation method, quantum electrodynamics, and statistical mechanics. 1965 edition, emended in 2005. 384pp. 6 1/8 x 9 1/4. 0-486-47722-3

Browse over 9,000 books at www.doverpublications.com

CATALOG OF DOVER BOOKS

Physics

INTRODUCTION TO MODERN OPTICS, Grant R. Fowles. A complete basic undergraduate course in modern optics for students in physics, technology, and engineering. The first half deals with classical physical optics; the second, quantum nature of light. Solutions. 336pp. 5 3/8 x 8 1/2. 0-486-65957-7

THE QUANTUM THEORY OF RADIATION: Third Edition, W. Heitler. The first comprehensive treatment of quantum physics in any language, this classic introduction to basic theory remains highly recommended and widely used, both as a text and as a reference. 1954 edition. 464pp. 5 3/8 x 8 1/2. 0-486-64558-4

QUANTUM FIELD THEORY, Claude Itzykson and Jean-Bernard Zuber. This comprehensive text begins with the standard quantization of electrodynamics and perturbative renormalization, advancing to functional methods, relativistic bound states, broken symmetries, nonabelian gauge fields, and asymptotic behavior. 1980 edition. 752pp. 6 1/2 x 9 1/4. 0-486-44568-2

FOUNDATIONS OF POTENTIAL THERY, Oliver D. Kellogg. Introduction to fundamentals of potential functions covers the force of gravity, fields of force, potentials, harmonic functions, electric images and Green's function, sequences of harmonic functions, fundamental existence theorems, and much more. 400pp. 5 3/8 x 8 1/2.
0-486-60144-7

FUNDAMENTALS OF MATHEMATICAL PHYSICS, Edgar A. Kraut. Indispensable for students of modern physics, this text provides the necessary background in mathematics to study the concepts of electromagnetic theory and quantum mechanics. 1967 edition. 480pp. 6 1/2 x 9 1/4. 0-486-45809-1

GEOMETRY AND LIGHT: The Science of Invisibility, Ulf Leonhardt and Thomas Philbin. Suitable for advanced undergraduate and graduate students of engineering, physics, and mathematics and scientific researchers of all types, this is the first authoritative text on invisibility and the science behind it. More than 100 full-color illustrations, plus exercises with solutions. 2010 edition. 288pp. 7 x 9 1/4. 0-486-47693-6

QUANTUM MECHANICS: New Approaches to Selected Topics, Harry J. Lipkin. Acclaimed as "excellent" (*Nature*) and "very original and refreshing" (*Physics Today*), these studies examine the Mössbauer effect, many-body quantum mechanics, scattering theory, Feynman diagrams, and relativistic quantum mechanics. 1973 edition. 480pp. 5 3/8 x 8 1/2. 0-486-45893-8

THEORY OF HEAT, James Clerk Maxwell. This classic sets forth the fundamentals of thermodynamics and kinetic theory simply enough to be understood by beginners, yet with enough subtlety to appeal to more advanced readers, too. 352pp. 5 3/8 x 8 1/2. 0-486-41735-2

QUANTUM MECHANICS, Albert Messiah. Subjects include formalism and its interpretation, analysis of simple systems, symmetries and invariance, methods of approximation, elements of relativistic quantum mechanics, much more. "Strongly recommended." – *American Journal of Physics*. 1152pp. 5 3/8 x 8 1/2. 0-486-40924-4

RELATIVISTIC QUANTUM FIELDS, Charles Nash. This graduate-level text contains techniques for performing calculations in quantum field theory. It focuses chiefly on the dimensional method and the renormalization group methods. Additional topics include functional integration and differentiation. 1978 edition. 240pp. 5 3/8 x 8 1/2.
0-486-47752-5

Browse over 9,000 books at www.doverpublications.com

CATALOG OF DOVER BOOKS

Physics

MATHEMATICAL TOOLS FOR PHYSICS, James Nearing. Encouraging students' development of intuition, this original work begins with a review of basic mathematics and advances to infinite series, complex algebra, differential equations, Fourier series, and more. 2010 edition. 496pp. 6 1/8 x 9 1/4. 0-486-48212-X

TREATISE ON THERMODYNAMICS, Max Planck. Great classic, still one of the best introductions to thermodynamics. Fundamentals, first and second principles of thermodynamics, applications to special states of equilibrium, more. Numerous worked examples. 1917 edition. 297pp. 5 3/8 x 8. 0-486-66371-X

AN INTRODUCTION TO RELATIVISTIC QUANTUM FIELD THEORY, Silvan S. Schweber. Complete, systematic, and self-contained, this text introduces modern quantum field theory. "Combines thorough knowledge with a high degree of didactic ability and a delightful style." – *Mathematical Reviews*. 1961 edition. 928pp. 5 3/8 x 8 1/2. 0-486-44228-4

THE ELECTROMAGNETIC FIELD, Albert Shadowitz. Comprehensive undergraduate text covers basics of electric and magnetic fields, building up to electromagnetic theory. Related topics include relativity theory. Over 900 problems, some with solutions. 1975 edition. 768pp. 5 5/8 x 8 1/4. 0-486-65660-8

THE PRINCIPLES OF STATISTICAL MECHANICS, Richard C. Tolman. Definitive treatise offers a concise exposition of classical statistical mechanics and a thorough elucidation of quantum statistical mechanics, plus applications of statistical mechanics to thermodynamic behavior. 1930 edition. 704pp. 5 5/8 x 8 1/4.
0-486-63896-0

INTRODUCTION TO THE PHYSICS OF FLUIDS AND SOLIDS, James S. Trefil. This interesting, informative survey by a well-known science author ranges from classical physics and geophysical topics, from the rings of Saturn and the rotation of the galaxy to underground nuclear tests. 1975 edition. 320pp. 5 3/8 x 8 1/2.
0-486-47437-2

STATISTICAL PHYSICS, Gregory H. Wannier. Classic text combines thermodynamics, statistical mechanics, and kinetic theory in one unified presentation. Topics include equilibrium statistics of special systems, kinetic theory, transport coefficients, and fluctuations. Problems with solutions. 1966 edition. 532pp. 5 3/8 x 8 1/2.
0-486-65401-X

SPACE, TIME, MATTER, Hermann Weyl. Excellent introduction probes deeply into Euclidean space, Riemann's space, Einstein's general relativity, gravitational waves and energy, and laws of conservation. "A classic of physics." – *British Journal for Philosophy and Science*. 330pp. 5 3/8 x 8 1/2. 0-486-60267-2

RANDOM VIBRATIONS: Theory and Practice, Paul H. Wirsching, Thomas L. Paez and Keith Ortiz. Comprehensive text and reference covers topics in probability, statistics, and random processes, plus methods for analyzing and controlling random vibrations. Suitable for graduate students and mechanical, structural, and aerospace engineers. 1995 edition. 464pp. 5 3/8 x 8 1/2. 0-486-45015-5

PHYSICS OF SHOCK WAVES AND HIGH-TEMPERATURE HYDRO DYNAMIC PHENOMENA, Ya B. Zel'dovich and Yu P. Raizer. Physical, chemical processes in gases at high temperatures are focus of outstanding text, which combines material from gas dynamics, shock-wave theory, thermodynamics and statistical physics, other fields. 284 illustrations. 1966–1967 edition. 944pp. 6 1/8 x 9 1/4.
0-486-42002-7

Browse over 9,000 books at www.doverpublications.com